INTERPRETING KUHN

Interpreting Kuhn provides a comprehensive, up-to-date study of Thomas Kuhn's philosophy and legacy. With twelve essays newly written by an international group of scholars, it covers a wide range of topics where Kuhn had an influence. Part I deals with foundational issues such as Kuhn's metaphysical assumptions, his relationship to Kant and Kantian philosophy, as well as contextual influences on his writing, including Cold War psychology and art. Part II tackles three Kuhnian concepts: normal science, incommensurability, and scientific revolutions. Part III deals with the Copernican Revolution in astronomy, the theory-ladenness of observation, scientific discovery, Kuhn's evolutionary analogies, and his theoretical monism. The volume is an ideal resource for advanced students seeking an overview of Kuhn's philosophy, and for specialists following the development of Kuhn scholarship.

K. BRAD WRAY is the author of *Kuhn's Evolutionary Social Epistemology* (Cambridge University Press, 2011) and *Resisting Scientific Realism* (Cambridge University Press, 2018). He has worked extensively in the Thomas Kuhn Archives at MIT. He teaches at the Center for Science Studies at Aarhus University.

INTERPRETING KUHN

Critical Essays

EDITED BY

K. BRAD WRAY

Aarhus University

CAMBRIDGE
UNIVERSITY PRESS

University Printing House, Cambridge CB2 8BS, United Kingdom

One Liberty Plaza, 20th Floor, New York, NY 10006, USA

477 Williamstown Road, Port Melbourne, VIC 3207, Australia

314–321, 3rd Floor, Plot 3, Splendor Forum, Jasola District Centre,
New Delhi – 110025, India

79 Anson Road, #06–04/06, Singapore 079906

Cambridge University Press is part of the University of Cambridge.

It furthers the University's mission by disseminating knowledge in the pursuit of education, learning, and research at the highest international levels of excellence.

www.cambridge.org
Information on this title: www.cambridge.org/9781108498296
DOI: 10.1017/9781108653206

© Cambridge University Press 2021

This publication is in copyright. Subject to statutory exception and to the provisions of relevant collective licensing agreements, no reproduction of any part may take place without the written permission of Cambridge University Press.

First published 2021

A catalogue record for this publication is available from the British Library.

ISBN 978-1-108-49829-6 Hardback

Cambridge University Press has no responsibility for the persistence or accuracy of URLs for external or third-party internet websites referred to in this publication and does not guarantee that any content on such websites is, or will remain, accurate or appropriate.

For Lori

Contents

List of Figures	*page* ix
List of Contributors	x
Acknowledgments	xi
List of Abbreviations	xii

    Introduction: The Road Ahead in Kuhn Scholarship   1
    K. Brad Wray

PART I FOUNDATIONAL ISSUES

1   The Genealogy of Thomas Kuhn's Metaphysics   9
    Paul Hoyningen-Huene

2   Kuhn's Kantian Dimensions   27
    Lydia Patton

3   A Public Intellectual and a Private Scholar:
    On Thomas Kuhn, James B. Conant, and the Place
    of History and Philosophy of Science in Postwar America   45
    George A. Reisch

4   Kuhn and Logical Positivism: On the Image of Science
    and the Image of Philosophy   65
    J. C. Pinto de Oliveira

PART II THREE CORE CONCEPTS

5   Mop-Up Work   85
    William Goodwin

6   Kuhn and the Varieties of Incommensurability   105
    William J. Devlin

7	Reassessing the Notion of a Kuhnian Revolution: What Happened in Twentieth-Century Chemistry *Eric R. Scerri*	125

PART III KUHNIAN THEMES

8	The Copernican Revolution since Kuhn *Peter Barker*	145
9	Kuhn, the Duck, and the Rabbit – Perception, Theory-Ladenness, and Creativity in Science *Vasso Kindi*	169
10	Kuhn on Scientific Discovery as Endogenous *Thomas Nickles*	185
11	Truth, Incoherence, and the Evolution of Science *Jouni-Matti Kuukkanen*	202
12	Reassessing Kuhn's Theoretical Monism: Addressing the Pluralists' Challenge *K. Brad Wray*	222

Bibliography	238
Index	264

Figures

7.1	Comparison of the extension of the term "planet" in the Ptolemaic theory and the Copernican theory	138
7.2	Venn diagram of the relationship between atomic number and atomic weight	140
8.1	Comparison of Ptolemy's, Tusi's, and Urdi's models for an outer planet	150
8.2	Four early European drawings of a Tusi couple	154

Contributors

PETER BARKER is Professor in the Department of History of Science at the University of Oklahoma.

WILLIAM J. DEVLIN is Professor in the Department of Philosophy at Bridgewater State University.

WILLIAM GOODWIN is Associate Professor in the Department of Philosophy at the University of South Florida.

PAUL HOYNINGEN-HUENE is Professor Emeritus at the Institute of Philosophy at Leibniz University, Hannover.

VASSO KINDI is Professor in the Department of History and Philosophy of Science at the National and Kapodistrian University of Athens.

JOUNI-MATTI KUUKKANEN is Associate Professor of Philosophy and founder and co-director at the Centre for Philosophical Studies of History at the University of Oulu.

THOMAS NICKLES is Foundation Professor Emeritus in the Department of Philosophy at the University of Nevada, Reno.

J. C. PINTO DE OLIVEIRA is Assistant Professor Emeritus in the Department of Philosophy at the State University of Campinas, Brazil.

LYDIA PATTON is Professor of Philosophy and Affiliate in the Department of Science, Technology and Society at Virginia Polytechnic Institute and State University.

GEORGE A. REISCH is an independent scholar, based in Chicago.

ERIC R. SCERRI teaches in the Department of Chemistry and Biochemistry at the University of California, Los Angeles.

K. BRAD WRAY is Associate Professor at the Centre for Science Studies at Aarhus University. He is a Life Member of Clare Hall, University of Cambridge.

Acknowledgments

I have many people to thank for helping me prepare this volume and see it into print. First, I thank Hilary Gaskin of Cambridge University Press. She first proposed such a volume to me in 2015 when I was on sabbatical, as a visiting scholar at MIT, completing my last book, *Resisting Scientific Realism*. I was able to begin work in earnest on this project, in 2018, after I moved to Denmark to take a position in the Centre for Science Studies at Aarhus University. Second, I would like to thank the contributors to the volume. I have known a number of them from the Kuhn scholarship community, but some I know only through their work. It has been a pleasure working with them. Third, I want to thank Aarhus University, and the Centre for Science Studies, in particular, for their support of my research. Fourth, I want to thank some of the scholars who have most shaped my own understanding of Kuhn and his work, specifically, Paul Hoyningen-Huene, Tom Nickles, and George Reisch. Fifth, I want to thank the Aarhus Universitets Forskningsfond for their support in the form of a Publication Grant – AUFF-F-2020–4-14. The grant has covered the costs related to the preparation of the index and proofreading.

Finally, I want to thank my partner, Lori Nash, to whom I dedicate this book. She has worked with me in preparing the manuscript for publication. In particular, she helped with editing some of the pieces, standardizing the citations to Kuhn's works, and preparing the footnotes and bibliography, all under quite tight time constraints. She also prepared the index and corrected the proofs. In addition, she has joined me in this adventurous move to Denmark, one of many moves I have made in order to pursue career opportunities. I acknowledge, and deeply appreciate, the sacrifices that she has made on this lifelong journey.

Abbreviations

BBT (1978). *Black-Body Theory and the Quantum Discontinuity, 1894–1912*. Oxford: Oxford University Press.

CR (1957/2003). *The Copernican Revolution: Planetary Astronomy in the Development of Western Thought*. Cambridge, MA: Harvard University Press.

ET (1977a). *The Essential Tension: Selected Studies in Scientific Tradition and Change*. Chicago: University of Chicago Press.

"Preface," pp. ix–xxiii, (1977b).
"The Relation between the History and the Philosophy of Science," pp. 3–20, (1976/1977).
"Concepts of Cause in the Development of Physics," pp. 21–30, (1971/1977).
"The Historical Structure of Scientific Discovery," pp. 165–177, (1962/1977).
"The Essential Tension: Tradition and Innovation in Scientific Research," pp. 225–239, (1959/1977).
"A Function for Thought Experiments," pp. 240–265, (1964/1977).
"Logic of Discovery or Psychology of Research?" pp. 266–292, (1970/1977).
"Second Thoughts on Paradigms," pp. 293–319, (1974/1977).
"Objectivity, Value Judgment, and Theory Choice," pp. 320–339, (1973/1977).
"Comment on the Relations of Science and Art," pp. 340–351, (1969/1977).

List of Abbreviations   xiii

PW   (n.d.) *The Plurality of Worlds: An Evolutionary Theory of Scientific Development.* Unpublished manuscript.

RSS   (2000a). *The Road since Structure: Philosophical Essays, 1970–1993, with an Autobiographical Interview,* eds. J. Conant and J. Haugeland. Chicago: Chicago University Press.

"What Are Scientific Revolutions?" pp. 13–32, (1987/2000).
"Commensurability, Comparability, Communicability," pp. 33–57, (1983/2000).
"Possible Worlds in History of Science," pp. 58–89, (1989/2000).
"The Road since *Structure*," pp. 90–104, (1991/2000).
"The Trouble with the Historical Philosophy of Science," pp. 105–120, (1992/2000).
"Reflections on My Critics," pp. 123–175, (1970/2000).
"Theory-Change as Structure-Change: Comments on the Sneed Formalism," pp. 176–195, (1976/2000).
"Metaphor in Science," 196–207, (1979/2000).
"Afterwords," pp. 224–252, (1993/2000).
"A Discussion with Thomas S. Kuhn" (with Aristides Baltas, Kostas Gavroglu, Vasso Kindi), pp. 253–323, (2000b).

SSR   *The Structure of Scientific Revolutions,* Chicago: University of Chicago Press.
First edition (1962a). **SSR-1**
Second edition, includes "Postscript-1969," (1962/1970). **SSR-2**
Third edition, includes "Postscript-1969," (1962/1996). **SSR-3**
Fourth edition, 50th anniversary edition. With an Introductory Essay by Ian Hacking (1962/2012), "Postscript-1969," pp. 173–208. **SSR-4**

TSK Archives – MC240	The following are documents from the Thomas Kuhn Archives at the Institute Archives and Special Collections, MIT Libraries, Cambridge, MA. They are cited according to date, box, and folder:
(1941).	"The War and My Crisis," box 1, folder 3.
(1943).	"Phi Beta Kappa Address," box 1, folder 3.

(1951a). Letter to Ralph Lowell, February 20, box 3, folder 10.
(1951b). Letter to Dean Owen, January 6, box 3, folder 10.
(1953). Letter to Charles Morris, July 3, box 25, folder 53.
(1955). "Can the Layman Know Science? State Teachers College – Bridgewater, Massachusetts 12/13/55," box 3, folder 33.
(1962b). Letter from Thomas S. Kuhn to Edwin B. Boring, November 29, box 4, folder 7.
(1967). "Paradigms and Theories in Scientific Research." box 3, folder 14. Cited from Marcum 2012a.
(1973). Letter to Kenneth Pietrzak, April 17, box 10.
(1984). "Lecture IV – Conveying the Past to the Present." Lectures/Meetings: Thalheimer Lectures, "Scientific Development and Lexical Change," box 23.
(n.d.) Untitled document ("Dear Professor Frank"), box 25, folder 53.
(n.d.) Kuhn M1: SSR – Chapter 1 – "What Are Scientific Revolutions?" box 4, folder 3.
(n.d.) Kuhn M2: SSR – Chapter 1 – "Discoveries as Revolutionary." box 4, folder 3.

No Abbreviations Used

The following are Kuhn's papers that do not appear in any of the above books. They are cited according to date.

(1945). "Subjective View. Thomas S. Kuhn, on Behalf of the Recent Students, Reflects Undergraduate Attitude." *Harvard Alumni Bulletin*, September 22, 29–30.
(1963). "The Function of Dogma in Scientific Research." In A. C. Crombie (ed.), *Scientific Change: Historical Studies in the Intellectual, Social and Technical Conditions for Scientific Discovery and Technical Innovation, from Antiquity to the Present*, pp. 347–369. London: Heinemann Educational Books.

(1969/1974).	"Second Thoughts on Paradigms -1969 Lecture." In F. Suppe (ed.), *The Structure of Scientific Theories*, pp. 459–482. Urbana, IL: University of Illinois Press.
(1979).	"Metaphor in Science". In A. Ortony (ed.), *Metaphor and Thought*, pp. 409–419. Cambridge: Cambridge University Press.
(1984/2017).	*Desarrollo científico y cambio de léxico. Conferencias Thalheimer*, ed. P. Melogno. Montevideo: ANII/UdelaR/SADAF.
(1990).	"The Road since Structure." In *PSA: Proceedings of the Biennial Meeting of the Philosophy of Science Association*, Vol. 1990, Volume Two: Symposia and Invited Papers, pp. 3–13.

Introduction
The Road Ahead in Kuhn Scholarship
K. Brad Wray

One might wonder if there is anything new to say about Thomas Kuhn and his views on science. Scholarship on Kuhn, though, has changed dramatically in the last twenty years. This is so for a number of reasons. Let me briefly mention three of them.

First, scholars studying Kuhn and his views are no longer focusing narrowly on Kuhn's *Structure of Scientific Revolutions* (SSR). Though this is undoubtedly Kuhn's most influential and most-read contribution to scholarship, it was not his final word on the topics discussed in the book. Scholars have been giving careful consideration to the papers in *The Road since Structure* (RSS), a collection of papers by Kuhn, written between the 1970s and the 1990s, published in 2000, four years after he died. These papers constitute Kuhn's final position on the philosophical topics about which he wrote, at least until his final unfinished book manuscript, *The Plurality of Worlds* (PW), is published. Already in the 1970s Kuhn was complaining that philosophers were still only addressing SSR. He was dismayed that scholars failed to account for what he had written since 1962. In many publications where Kuhn's work is discussed, especially those directed at a more general audience, the focus still tends to be exclusively on SSR (see, e.g., Morris 2018). Now, though, it is common to discuss Kuhn's mature views, at least among more serious scholars. The papers published in RSS provide clarification of Kuhn's earlier account, as well as modifications in his thinking, thus allowing us to develop a more historically sensitive account of Kuhn's views.

Second, many scholars, including a number of contributors to this volume, have been drawing on the vast unpublished resources at the Thomas S. Kuhn Archives, in the Special Collections Library at the Massachusetts Institute of Technology (TSK Archives-MC240). The twenty-six or so boxes Kuhn left to MIT upon his retirement include draft presentations, lecture notes from courses he gave throughout his career, correspondence with scholarly colleagues, fan mail, early notebooks, and much more (see Wray 2018b). Indeed, there is now an active informal network among those who have conducted research at MIT, where notes are shared, and attention is drawn to hitherto neglected documents that have helped us develop a richer and more complete picture of Thomas Kuhn, the person and scholar. George Reisch and Paul Hoyningen-Huene, in particular, have been especially important in bringing these resources to light (see, e.g., Hoyningen-Huene 2015; Reisch 2016; also Reisch 2019a).

Third, with the fiftieth anniversary of the publication of SSR in 2012, there were quite a number of conferences held. This led to the publication of a number of volumes reflecting on Kuhn's impact in philosophy and history of science (see, e.g., Blum, Gavroglu, Joas, and Renn 2016; Devlin and Bokulich 2015; Kindi and Arabatzis 2012; and Richards and Daston 2016). These volumes included contributions from both leaders in the field who had worked or studied with Kuhn (see, e.g., Heilbron 2016), as well as those who worked in the years when Kuhn's influence was ubiquitous and at its peak (see, e.g., Bloor 2016).

These three developments have contributed significantly to our collective understanding of Kuhn and his theories of scientific change and scientific knowledge.

Kuhn's position in the philosophy of science can be difficult to gauge at times. For every enthusiastic scholar who appreciates the insight Kuhn provides, there is another who insists that he had an adverse effect on our understanding of science and that his impact is exaggerated. But by objective measures, Kuhn's impact is undeniable. In an extensive bibliometric study, Wray and Bornmann found that Kuhn was responsible for two of the seven citation peaks in philosophy of science between 1900 and 1970. Wray and Bornmann examined a huge data set of citations in key philosophy of science journals in an effort to identify citation peaks, the years in which highly cited sources in philosophy of science were published. One peak occurred in 1962, the year when SSR was first published, and another occurred in 1970, the year the second edition with its Postscript (SSR-2) was published (see Wray and Bornmann 2015). In fact, normalizing the citation curve for the whole period, SSR is responsible for the two highest citation peaks in the period from 1900 to 1970. Importantly, Wray and Bornmann limited their search of citations to key journals in the philosophy of science: *Erkenntnis*, *Philosophy of Science*, *British Journal for the Philosophy of Science*, and *Studies in History and Philosophy of Science*. So these peaks are not a consequence of the vast number of citations that Kuhn's SSR receives from scholars in other fields.

Introduction: The Road Ahead in Kuhn Scholarship

Kuhn's influence outside of philosophy of science is also astounding, especially in the social sciences. In 2016, Elliott Green reported that Kuhn's book SSR was the most-cited book in the social sciences (see Green 2016), even more cited than Michel Foucault's *Discipline and Punish*, and *The History of Sexuality*, and Karl Marx's *Das Kapital*. Eugene Garfield found that Kuhn was the forty-fourth most-cited author in the *Arts and Humanities Citation Index* for 1977 and 1978 (see Garfield 1979–80, 240). In a more recent study of most-cited authors of books in the humanities, looking at citations from 2007, Kuhn ranked thirty-fifth (see THE 2009). So it is beyond dispute that Kuhn has had a profound and wide-ranging impact on scholarship.

A title like *Interpreting Kuhn* invites a wide range of responses. In fact, the contributors to this volume have explored a range of aspects of Kuhn's philosophy of science. The various contributions are collected into three parts. Part I is concerned with foundational issues in Kuhn scholarship. The essays in this part explore Kuhn's fundamental assumptions, his metaphysical assumptions, and his relationship to Kant, as well as influences on Kuhn, both conceptual and contextual. Part II is focused on key Kuhnian concepts. Much of the appeal of SSR was due to the engaging concepts and metaphors that Kuhn developed and employed in his analyses. Some of them, for example, incommensurability, have given rise to lively scholarly debates and a vast body of scholarly literature. Others, like normal science and scientific revolutions, have become central terms in our understanding of science, even if there is still no consensus on what these terms denote. Part III contains essays exploring various themes that run through Kuhn's work. This part includes reflections on the Copernican Revolution in astronomy since the publication of Kuhn's *The Copernican Revolution: Planetary Astronomy in the Development of Western Thought* (CR), as well as reflections on the nature of scientific discovery and the theory-ladenness of scientific observation. There are also essays on the evolutionary dimensions in Kuhn's philosophy, as well as theoretical monism, themes that rose out of Kuhn's work in philosophy of science.

I want to highlight some of the specific themes discussed in the various contributions.

Part I is devoted to foundational issues. Paul Hoyningen-Huene analyzes the metaphysics underlying Kuhn's view. Hoyningen-Huene argues that Kuhn's work should be understood as part of a larger project, extending at least as far back as Copernicus, and developed further by Kant, which aims to emphasize the subject-sided contributions to our understanding of the world. In a related vein, Lydia Patton explores Kantian themes in

Kuhn's work. She argues that though both Kuhn and Kant were concerned with the question of the status of science, scientific communities figure centrally in Kuhn's analysis, but not Kant's analysis. George Reisch examines Kuhn's relationship to both James B. Conant, his mentor at Harvard, and the Cold War culture in America. J. C. Pinto de Oliveira examines the influence of art on Kuhn's thinking when he was writing SSR. As he shows, earlier manuscript versions of *Structure* contained comparisons between art and science, many of which were not retained in the final manuscript.

Part II is concerned with key Kuhnian concepts: normal science, incommensurability, and scientific revolutions. William Goodwin provides a much-needed comprehensive analysis of normal science, that is, the process Kuhn described as mopping-up. Normal science has tended to be in the shadow of the far more exciting concept, scientific revolution, despite the fact that Kuhn believed that most scientists, most of the time, are engaged in normal science. William Devlin provides a useful analysis of incommensurability, a concept that turns out to be multifaceted, though not quite as multifaceted as the paradigm concept. Devlin argues that all forms of incommensurability that figure in Kuhn's work are incompatible with the correspondence theory of truth. And Eric Scerri provides an analysis of the notion of a Kuhnian revolution, through a detailed examination of an episode in the history of chemistry, when chemists came to classify chemical elements by their atomic number, rather than their atomic weight. Scerri considers the extent to which this change in chemistry constitutes a revolution in the sense articulated in SSR and in the sense articulated in Kuhn's later works, where Kuhn invokes the notion of a lexical change.

One might wonder why there is no contribution in this volume dedicated to the paradigm concept. The paradigm concept has been written about and analyzed extensively, and Kuhn distanced himself from the concept in the early to mid-1970s (see Wray 2011, chapter 3). Thereafter, Kuhn would refer to the concrete scientific achievements that set a scientific field on the developmental cycle, which he describes in SSR, as exemplars. And exemplars were distinguished from both (i) theories and (ii) disciplinary matrices, two notions that he indiscriminately referred to in SSR as paradigms as well. Kuhn acknowledged that he overused the term "paradigm." In the 1970s, when he published *Essential Tension* (ET), he remarked that "challenged to explain the absence of an index, I regularly point out that its most frequently consulted entry would be: 'paradigm, 1–172, passim'" (294). Although Kuhn came to regret his rather undisciplined use of paradigm, he retained a sense of humor about it. When he was asked about how he came up with the term in an interview in the 1990s, he remarked that "*paradigm* was a perfectly good word, until I messed it up" (see RSS, 298). Still, the paradigm concept is discussed in a number of the contributions to this volume.

Part III is devoted to Kuhnian themes and explorations of Kuhn's relevance to contemporary scholarship. Peter Barker provides a provocative analysis of the Copernican Revolution in light of the decades of scholarship since the publication of Kuhn's first book, CR. He also reconsiders the notion of a Copernican Revolution in astronomy. Vasso Kindi examines the relationship between theory-ladenness and perception in Kuhn's work, tracing the significance of Gestalt psychology to Kuhn's thinking about science. In particular, Kindi emphasizes the importance of the duck-rabbit figure to Kuhn in his efforts to explicate the notion of a scientific revolution. Tom Nickles provides an analysis of the logic and structure of discovery in science, a theme that was quite important to Kuhn. Nickles draws attention to a tension in Kuhn's work. In making discovery a normal part of normal science, he seems to leave unexplained the origins of the sorts of radical discoveries that cause revolutionary changes of theory. Jouni-Matti Kuukkanen examines the role of truth in Kuhn's account of science, with special attention to its relationship to the evolutionary dimensions in Kuhn's writings. Returning to a theme discussed by Devlin, Kuukkanen argues that Kuhn's account of science can be adequately served by a pragmatic notion of truth. Finally, K. Brad Wray examines the apparent threat posed to Kuhn's theoretical monism by recent developments in philosophy of science that emphasize the value and role of pluralism. Special attention is given to Hasok Chang's influential work on pluralism.

These essays will provide valuable insights into Kuhn's views and should prove useful not only to scholars in philosophy, but in other fields as well.

PART I

Foundational Issues

CHAPTER 1

The Genealogy of Thomas Kuhn's Metaphysics
Paul Hoyningen-Huene

1.1 Introduction

In his *Structure of Scientific Revolutions* (SSR) of 1962, Thomas Kuhn famously stated: "In a sense that I am unable to explicate further, the proponents of competing paradigms practice their trade in different worlds" (SSR-2, p. 150).

More than twenty years later, Kuhn came back to this topic and asked: "Is this an idealist position? Perhaps it is. But the idealism is then unlike any other of which I am aware" (Kuhn 1984, p. 122).

Probably for most of his readers who are critical of that position, the novelty of his idealism will not be true solace. Also, the following sentence will not change the critical readers' mind: "Perhaps it is an idealist's world nonetheless, but it feels very real to me" (Kuhn 1984, p. 123).

In fact, quite soon after the publication of SSR, Israel Scheffler reacted in a way that was highly influential for the further philosophical reception of Kuhn's position: "I cannot, myself, believe that this bleak picture, representing an extravagant idealism, is true" (Scheffler 1967, p. 19).

The problem seems to be that realists do not understand how any idealism can present the world – that is seen by idealism as mind-dependent – as real.

In this chapter, I will not really argue for Kuhn's metaphysical position. An argument only makes sense if one knows what one is arguing for. Kuhn's metaphysical position is, however, rather elusive, of which Kuhn was aware. Therefore, I will try to make Kuhn's position better intelligible. Perhaps I will also be able to contribute some initial plausibility, similar to the "reasonable suspicion" that justifies further investigation of a legal case. For that purpose, I shall put Kuhn's attempted metaphysical position into a larger historical context, its genealogy. I shall begin with two developments that proved of utter significance for Western thought: first, the Copernican revolution

(Section 1.2) and, second, the distinction between primary and secondary qualities (Section 1.3). Both items figure fundamentally in the emergence of modern physics. The Copernican revolution especially played a major role in Kant's critical philosophy, because it transported the specific post-Copernican mode of thought into philosophy. However, the distinction between primary and secondary qualities is also relevant in this context, for in order to make the idealist components of Kant's critical philosophy more digestible, it may be formulated in terms of that distinction (Section 1.4). The next step in this genealogy will be twentieth-century physics, because in both special relativity theory and quantum theory, a post-Copernican element can be found (Section 1.5). Finally, we reach Kuhn and can, in the light of the earlier developments, articulate his position as belonging to the given genealogy (Section 1.6). My hope is that this will make Kuhn's attempted metaphysical position more accessible, especially to those who, when reading "world change" and similar expressions, quickly turn away in hardly veiled contempt.[1]

1.2 Copernicus

The starting point of the genealogy of Kuhn's metaphysics is the Copernican theory of the planetary system.[2] Copernicus suggested that the Earth is not at the center of the planetary system, but the Sun is. This has important consequences for our understanding of the observable motions of all celestial bodies. For my purposes, it is sufficient to discuss the motion of the Sun. In the geocentric view, the motion of the Sun was real and objective: It was causally efficacious, and a property solely of the Sun. There was no difference between the objectivity and reality status of, for instance, the radiation of the Sun and the Sun's motion in the sky; both were seen as undoubtedly real and causally efficacious.

In some sense, nothing in the causal efficacy of the motion of the Sun has changed in the meantime. For instance, the motion of the Sun generates the difference between day and night, and not moving your sunshade according to the Sun's motion may cause a real sunburn, which may not be an apparent one. It seems that producing causal effects is a *sufficient*

---

[1] For a recent attempt to defend Kuhn's world change talk by a historical case study, see Wray and Andersen 2019.
[2] I am giving here an extremely simplified version of the Copernican theory that abstracts from everything that is not relevant in our context. For more responsible versions that include aspects of the complicated historical development of the theory, see, e.g., Andersen, Barker, and Chen 2006, pp. 130–163; CR.

indicator of something being real.³ It also seems impossible that something unreal, that is, not existent, can generate causal effects. The geocentric worldview was just completely natural in the sense that it identified something in the heavens that had effects on the Earth as real, including the motion of the Sun. This is also in full accordance with a general stance of ancient and medieval philosophy and science to see the real as completely object-sided, as independent and devoid of any components that have their origin on the subject side. This is, of course, also our everyday view of reality: Real in the relevant sense⁴ are the things that stand opposite to us, that are completely independent of us, and that have, as I shall express it, no components that have their origin at the subject side ("genetically subject-sided components" for short; "genetically" does not refer to "genetics," but to "genesis"). Thus, in our common understanding we speak very naturally of reality as "mind-independent."

In this view, the following equation holds:

> real = objective = purely object-sided = without genetically subject-sided components.⁵

Given this view of reality, it is completely intelligible and plausible that any form of "idealism" in metaphysics appears to be incoherent and thus not even worth discussing. Most generally, any form of idealism seems to claim that something genetically subject-sided is at least part of, or even fully constitutes, reality.⁶ To someone committed to the above view of reality, this is incoherent, because by "reality" we just *mean* the completely object-sided, to which the epistemic subject does not and cannot contribute anything whatsoever.

---

³ The close connection between reality and causal efficacy has been called the "Eleatic principle" by David Armstrong, following a passage in Plato's *Sophist*; see, e.g., Colyvan 1998, p. 313 with fn. 1. I wish to thank Howard Sankey for making me aware of the expression "Eleatic principle."

⁴ The relevant sense here covers things that are located outside of us. Of course, also in our everyday understanding we know about the reality of events or processes that are located in our consciousness, like feelings. In the text, I am only referring to external reality because the fundamental subject here is the natural sciences.

⁵ Note especially that in this equation "objective" and "purely object-sided" coincide. This is a consequence of the fact that our language is adapted to (commonsense) realism. The attempt to articulate philosophical positions different from realism therefore meets linguistic difficulties, which is why I distinguish "objective" from "purely object-sided"; otherwise one can only speak in paradoxes. I have introduced and used the terms "subject-sided" and "object-sided" earlier in Hoyningen-Huene 1993, pp. 33–36, 45–47, 62–66, 122 fn 283, 125, 267–271; Hoyningen-Huene, Oberheim, and Andersen 1996, p. 139; Hoyningen-Huene and Oberheim 2009, p. 208.

⁶ I am aware of the fact that "absolute" or "objective" idealists would not agree with this characterization of idealism because they would deny that the respective entity, spirit or nous or whatever is in any useful sense subjective. This is, however, not important in our context.

Now, Copernicus claims that the Sun's observed motion in the sky is an "apparent" motion in contrast to the "true" (or "real" or "own") motion. In fact, as the Sun is at the center of the planetary system, it may be said that the Sun is at rest. The daily apparent motion of the Sun from sunrise in the East to sunset in the West is generated by the Earth's rotation and, because we cannot sense the Earth's rotation, by the projection of this rotation onto the sky. More generally, the motions of all celestial bodies are "apparent motions," consisting of two components. One component is the true motion, for instance, in the case of the planets, their motion around the Sun. The second component is added to the first one. It results from our point of view on a rotating and Sun-orbiting Earth. The addition of these motion components results in the apparent motion of the respective celestial body, that is, the motion of the celestial body as it appears to us observers on the Earth. The case of the Sun is particularly simple because the Sun's own motion is zero, so in the course of a day we only see the projection of the Earth's rotation onto the sky, which results in an (approximately) uniform motion of the Sun in the sky. In the case of the planets, however, their apparent motion is at times quite unexpected, namely retrograde. In antiquity, this gave rise to complicated constructions of epicycles and various other devices in order to "save the phenomena," that is, to account for the supposedly real motions of the planets.

Now, it seems completely natural to call the observed motion of the Sun and of the other celestial bodies "objective." There is no phenomenological difference between the objectivity of the observed motion of a horse that passes by and the observed motion of the Sun or the Moon in the sky. Different observers agree on these motions, these motions are thus intersubjectively observable, and they can be consistently measured by physically very different devices to very high degrees of accuracy. All this contributes to the impression that these motions are independent of any contributions by the observer and thus objective and real.

However, under a Copernican analysis, the observed motion of a planet and of other celestial bodies contains both genetically *object*-sided contributions (the planet's own, "true" motion) and genetically *subject*-sided contributions (the projection of the Earth's rotation and motion onto the planet's observed motion). Several features of this situation are noteworthy.

First, the observable motion of a celestial body is a *phenomenologically* inextricable mixture of the projection of Earth's rotation and motion, and the true motion of the respective body. For instance, no analysis of a planet's trajectory alone can reveal *that* there are genetically subject-sided contributions involved, let alone distinguish the genetically subject-sided components from the genetically object-sided ones.

Second, the phenomenological indistinguishability of genetically subject-sided and object-sided contributions to the apparent motion is reflected in classical physics by the fact that they can be treated mathematically in identical ways. In the calculation of the apparent motion, the subject-sided contribution is vector-added to the object-sided contribution in the same way as two purely object-sided motion components would be vector-added.[7]

Third, apparent motions are no indicators whatsoever of the true motions; in fact, taken singly, they are completely misleading about the true motion. The reason is that the observer's motion is entirely independent of the true motion of the object, and it can overwhelm the true motion component of the apparent motion. For instance, in the case of the Sun, its apparent motion across the sky is absolutely no indicator of its true motion, namely zero motion. (This does of course not exclude the possibility to calculate the true motion once the apparent motion and the observer's motion are known.)

Fourth, the phenomenological equivalence of genetically subject-sided and object-sided contributions to apparent motions immediately extends to its causal effects. In order to avoid a sunburn, one has to move one's sunshade, independently of whether the Sun's observed motion is, in the Copernican sense, only apparent or true. This is why the geocentric planetary system was historically so stable. Only if one puts the planetary motions into a larger theoretical context and adduces data *beyond* planetary position data does the superiority of the heliocentric system become intelligible.

The Copernican system clearly has had massive substantial consequences for planetary astronomy. I believe, however, that its methodological consequences exceed the substantial ones by far. In fact, arguably the Copernican insights represent the most important turning point in the history of Western scientific and philosophical thought. In antiquity and the Middle Ages, the way natural phenomena presented themselves as purely object-sided was basically accepted as veridical. Of course, our senses may deceive us from time to time, and the phenomena presented to us may not be the ultimate reality, as most famously Plato thought.[8] And

---

[7] The deeper reason for this symmetrical treatment of a projected motion and the true motion is that Newtonian physics is Galilei invariant.
[8] Here is a fundamental difference between Plato's and Copernicus's figures of thought. In his allegory of the cave, Plato distinguishes between shadows – the analogue to the phenomena we have access to – and true things (ideas), to which we have no access. In this setting, the shadows are at least indicators of true things, even if the shadows do not represent them truthfully. The analogue does not hold in the Copernican case.

of course, Aristarchus of Samos already in antiquity considered the possibility that the Sun is at the center of the planetary system. But by and large, natural phenomena were seen as purely object-sided, as our commonsense realism also suggests.

Now the fundamentally disruptive insight of Copernicanism is that real, objective, causally efficacious phenomena that *appear* to be purely object-sided *are not* necessarily purely object-sided, because they may contain genetically subject-sided components. Whether that is true of a particular case and what the genetically subject-sided components possibly are cannot be discovered by investigating the phenomenal qualities of the phenomenon alone, but only by putting it into a larger theoretical context. In other words, what is real and objective *and thus* appears to be purely object-sided may in truth contain genetically subject-sided contributions, that is, components that in some sense stem from us. And most importantly, this is not just a *philosophical* speculation, but a *scientific* insight that proved immensely successful in its application to planetary theory in the Copernican system. Without considering the possibility that objectively determinable orbits of celestial bodies may nevertheless contain genetically subject-sided contributions, the Copernican system could not have been invented. The identification of genetically subject-sided components in some set of phenomena is scientifically of utmost importance, because they have to be investigated separately from and in a different way than the genetically object-sided contributions.

The analysis of an *earthly* phenomenon may be useful to further illustrate the properties of the planetary case. In 1851, French physicist Léon Foucault mounted a 67-meter-long pendulum in the dome of the Panthéon in Paris. Once the pendulum is set in motion, the plane of its motion does not stay stationary, as one would expect from everyday experience or from the law of the conservation of momentum. In fact, the plane of the pendulum's motion rotates due to the Earth's rotation; relative to the Sun, it stays stable due to conservation of momentum. This motion is causally efficacious, as is demonstrated by many installations of Foucault's pendulum, where the pendulum displaces small objects once they are reached by the rotating plane of motion and the pendulum knocks them over.[9] Again, this is a phenomenon fulfilling all criteria of objectivity, and still, it has massively genetically subject-sided components.

---

[9] For a demonstration of the effect, see, e.g., www.youtube.com/watch?v=iqpV1236_Q0 (accessed September 25, 2019).

The consequences of Copernicus's discovery of the difference between apparent and true motions have had dramatic consequences for Western scientific and philosophical thought, until today, because it conveys a fundamental lesson that has been taken very seriously in science ever since. Any phenomenon fulfilling all tests of its reality such as object-sidedness, resistance, robustness, and causal efficacy may still contain genetically subject-sided components that cannot be discovered by analyzing the qualities of the phenomenon itself. These genetically subject-sided components may come to the fore if the phenomenon is embedded in a larger theoretical context. Most importantly, this insight was not the result of philosophical speculation but was the essential ingredient of a fundamental and extremely successful revolution in astronomy, which was ultimately accepted because of the pressure of empirical data. The same critical thought that was at the bottom of Copernicus's dramatic innovation was also at the bottom of the equally dramatic innovation of physics in the seventeenth century. Galilei, Boyle, Descartes, and others, who were the fathers of this revolution, made a distinction similar to the Copernican distinction between apparent and real motion. It rested on the same insight: that something apparently purely object-sided and objective may in truth contain genetically subject-sided components.

1.3 Primary and Secondary Qualities

As the Copernican planetary system had its predecessors in antiquity, so was the distinction between primary and secondary qualities already made in antiquity by the atomists Leucippus and Democritus.[10] In comparison to the philosophical thought of Plato and Aristotle, their effect on the development of science and philosophy was firstly rather meager. Because in the beginning of the modern phase of science corpuscularism was the dominant metaphysics, problems and answers with some similarity to their ancient analogues emerged again. All the main contributors to the emergence of modern physics such as Galilei, Descartes, Hobbes, Boyle, Locke, and Newton discussed and used some version of the distinction between primary and secondary qualities.[11] For the purpose of this chapter, I do not have to consider different variants

---

[10] The historical situation is, in fact, much more complicated; see, e.g., Lee 2011.
[11] See Alexander 1974; Ayers 2011; Burtt 1932/1980, pp. 83–90, 115–121, 130–134, 180–184, 231–239; Campbell 1980; Curley 1972; Keating 1993; Martinez 1974; McCann 2011; Palmer 1976.

because they roughly agree about what I need here as their fundamental aspect.[12]

Primary qualities were thought of as inherent qualities of the respective object. In the seventeenth century, the mostly agreed-upon examples were the object's geometrical shape, its size, or its motion. By contrast, secondary qualities of an object were qualities that resulted from the interaction of the object with our senses. The most agreed-upon examples were the object's taste, odor, color, sound, or heat. As in Copernicus's case of apparent motions, secondary qualities are no substantive indicators of the content of primary qualities. This is because primary and secondary qualities stand in many-many relations: The same set of primary qualities can generate different secondary qualities in different circumstances, or in different observers, and the same secondary quality can be generated by different sets of primary qualities. For instance, the same phenomenological red can be generated by light consisting of one wavelength or of various mixtures of different wavelengths. Thus, it is impossible to infer from some secondary quality alone anything about the underlying primary qualities.

The fundamental importance of the distinction between primary and secondary qualities derives from the fact that primary qualities were the subject matter of the emerging mathematical physics, whereas secondary qualities were not. Thus, the delineation of the subject matter of the emerging modern physics was given by the distinction between primary and secondary qualities. The distinction in a sense neutralized those qualities that did not seem to be amenable to a mathematical treatment, that is, the secondary qualities. The distinction pushed them, together with their messiness in comparison with the mathematical crispness of the primary qualities, out of the domain of physics.

Put in the formerly used terminology, primary qualities were seen as purely object-sided, whereas secondary qualities as containing genetically subject-sided components as well. Historically, however, there were differences of opinion regarding what criterion would identify primary qualities, how exactly to conceive of secondary qualities, and what the reality status of secondary qualities was. What is utterly important in our context, however, is the fundamental parallel between the distinctions of primary versus secondary qualities and true motions versus apparent motions. I found only one author who has seen this parallel and their historical

---

[12] For general discussion of the primary–secondary distinction, see, e.g., Averill 1982; Hirst 1967; Macintosh 1976; Smith 1999; Vision 1982.

connection, namely Edwin Arthur Burtt in his influential *The Metaphysical Foundations of Modern Physical Science* of 1932:

> The secondary qualities are declared to be effects on the senses of the primary qualities which are alone real in nature. [. . .] This doctrine, too, was bolstered up by considerations derived from the Copernican astronomy. Just as the deceptive appearance of the earth, which makes us suppose it to be at rest, arises from the position and local motion of the onlooker, so these deceptive secondary qualities arise from the fact that our knowledge of objects is mediated by the senses. (Burtt 1932/1980, p. 84)

In both cases, the pertinent phenomena or qualities, only later dubbed "apparent" or "secondary," appeared prior to that to be purely object-sided, because of their phenomenological qualities and their causal efficaciousness, as could be (quasi-)experimentally demonstrated. To appreciate the causal efficacy of the later so-called secondary qualities, consider, for example, the loud bang of an explosion that may damage the eardrum; odors that influence the behavior of animals and humans; the light of a specific color that triggers a particular photochemical reaction (a Whiggish example). To appreciate the causal efficacy of the later so-called apparent motions, consider the difference between night and day, which is caused by the Sun's motion.

Consistent with their causal efficacy, in antiquity and the Middle Ages, later so-called secondary qualities and apparent motions were mostly seen as real: Their reality status was on equal footing with other unquestionably real properties.[13] Note that the same still holds for common sense (see, e.g., Chirimuuta 2015, pp. 31–37). For instance, in everyday life the material and the color of a sweater have the same reality status. In the same vein, in ordinary language we call certain people "colorblind," as if colors were something really existing outside of human beings in the real world and that these people are just incapable of grasping them.

It is interesting to see how in the Scientific Revolution, after their downgrading to secondary qualities or apparent motions, respectively, the formerly unquestionably real properties and motions changed their reality status.[14] This is because they are now seen as having lost their genuine causal powers. The general mechanism of this process is this. As soon as certain qualities or motions are seen as secondary or as only

---

[13] For colors, see, e.g., Chirimuuta 2015, pp. 19–22, 49–52.
[14] This holds for most authors, but not for Boyle, who insisted on the full reality status of secondary qualities; see Burtt 1932/1980, pp. 180–184.

apparent, respectively, they are no longer attributed to the respective objects themselves but are primarily located in the perceiving subject. Secondary qualities and apparent motions are the effects of the primary qualities and the true motions, respectively, on the perceiving subject. According to this view, the causal efficacy is now relocated, away from secondary qualities and apparent motions, to the underlying primary qualities and true motions. For example, it is not the loud bang of an explosion, a secondary quality, that damages the eardrum, but the underlying violent motion of the air molecules, a primary quality. Or it is not the color of the light that triggers a particular photochemical reaction, but the frequency-dependent energy of the underlying electromagnetic waves. Similarly, for motions, it is not the Sun's apparent motion that causes day and night, but the rotation of the Earth. Or it is not Foucault's pendulum's own motion alone that produces the collision with the little objects on the periphery of the swinging pendulum, but the conjunction of the pendulum's own motion with the contribution by the Earth's rotation.

The lesson of these groundbreaking changes that are essential elements of the transition from the ancient and medieval conception of science to modern science is this. It is illegitimate to infer from the *phenomenological* quality of a phenomenon as purely object-sided and as causally efficacious that it *really* is purely object-sided and is indeed itself causally efficacious. The phenomenon in question may contain genetically subject-sided components that are phenomenologically *completely* hidden. Note that this lesson is not the result of some philosophical speculation, or of skepticism,[15] but is the result of one of the greatest events in human intellectual history, the transition from medieval to modern science. It is this insight that is, together with other elements, at the bottom of modern science. Disregarding this insight means nothing less than denying the immeasurable progress that scientific thinking has made due to the scientific revolution. Disregarding this insight also means staying in a pre-Copernican scheme of thought.

For example, G. E. Moore's famous defense of realism by showing his hand and exclaiming "this is a human hand," as an example of a real object, fails because it tries to work from the phenomenological quality of his hand *alone* to its pure object-sidedness (see Moore 1925). This is like doubting

---

[15] I am mentioning skepticism here because often realists try to fend off arguments against realism as parts of skepticism, and as skepticism is not seen as a defensible position, the respective arguments are dismissed.

the Earth's rotation because it cannot be felt or claiming the absolute reality of the Sun's motion because it can be observed. Putting these phenomena into a greater theoretical context may change their reality status. Again, this is not skepticism; it is post-Copernican thinking.

Here is the positive result of these considerations for both science and philosophy. Whenever plausible, the possibility of phenomenologically imperceptible genetically subject-sided components of something that presents itself as purely object-sided must be considered. And this is exactly what has happened since the seventeenth century in both science and philosophy. For instance, significant parts of theoretical philosophy from the seventeenth until at least the mid-nineteenth century can be seen as controversies about the identification of genetically subject-sided components in seemingly purely object-sided phenomena: Where exactly is the dividing line between the genetically subject-sided and the purely object-sided?[16]

As I claim that this part of post-Copernican thinking is essential in the genealogy of Kuhn's metaphysics, I shall in the following very briefly sketch some historical cases where post-Copernican thinking became operative. These cases will differ with respect to their historical success, as far as we can assess it today.

1.4 Kant

Kant himself is very explicit about the line that connects his critical philosophy with the Copernican revolution. He claims that his own "Copernican turn" in philosophy is modeled upon what Copernicus did for planetary theory.[17] This means for Kant: Try to decipher phenomena that appear to be purely object-sided as having phenomenologically hidden genetically subject-sided components. For mathematics, explain the possibility of mathematical proofs by the genetical subject-sidedness of time and space. For physics, explain the existing a priori regularities in nature by their origin from the subject side. One source of inspiration for the latter is David Hume, who according to Kant's own judgement, awoke him from his "dogmatic slumber." The slumber consisted in the rationalist heritage firstly to assume causality to be purely object-sided and secondly as

---

[16] Arthur Schopenhauer has given a schematic description of the history of philosophy from Descartes to Kant along these lines; see Schopenhauer 1851/2014.
[17] See Kant (1781&1787/1998), ed., tr. Guyer, and Wood, pp. B XVI and especially B XXII fn. *. In the Kant literature, it was especially the German Kant scholar Friedrich Kaulbach who stressed the importance of the "Copernican thinking figure"; (see Kaulbach 1973).

epistemically accessible to pure nonempirical thought. Instead, Hume attributed causality to the subject side and made it somehow "subjective." Kant thought that by his transcendental philosophy, he was able to agree with Hume's attribution of causality to the subject side but at the same time to restore its objective and a priori character.

Kant maintains that *all* phenomena accessible for us contain genetically subject-sided components. He also expresses this in the terminology of primary and secondary qualities: *All* physical qualities are secondary qualities, including especially space, time, and causality.[18] According to Kant, the contributions to apparently purely object-sided phenomena from the subject side are individually and historically invariable: They are the "forms" of intuition and thought. These forms are universal for human beings, and they are necessary preconditions for any experience. As our only way to get in contact with the world is through these forms, we have no access whatsoever to the primary qualities that underlie the manifest secondary qualities of real phenomena.

It is extremely important that in Kant's view, physical *reality* consists of what he calls appearances, and there is no other reality for us: Physical reality consists of causally interacting material things. *All physical reality, that is, all material things, contains genetically subject-sided elements, which do not, however, diminish in the least their status as real things.* With respect to the properties of real things, we are thus in a state like the state of the ancients with respect to colors. Colors were successfully taken as real, with causal forces, and no phenomenological investigation into colors could throw doubt on their reality. It was a disruptive change of perspective in the seventeenth century due to a new theoretical system that diminished their reality status to that of secondary qualities and transferred their causal powers to the underlying primary qualities. Kant, however, objected that these putative primary qualities are, in fact, also secondary qualities with the fundamental difference that in principle, there is no theoretical system available that allows us access to the underlying truly primary qualities. Reality is this set of causally interacting material things, and their genetically subject-sided components do not change anything in their reality.

My point in this section is not to defend Kant's position. It seems to me that today, very little of substance can indeed be defended of it. The main reason is that Kant built his position by using parts of the logical, mathematical, and scientific knowledge of his day, which he thought were

---

[18] Kant (1783/2004), ed., tr. Hatfield, 4:289; for discussion, see Allais 2007; Rosefeldt 2007; Allais 2015, especially pp. 125–144.

eternal. As it turned out in the nineteenth and twentieth centuries, nothing of these putative eternal truths survived unscathed; in fact, some of them were virtually abolished. But Kant illustrates how post-Copernican thinking made its way further into philosophy. We shall see in a moment that it also made its way into some of the most breathtaking innovations of twentieth-century physics.

In addition, due to Kant's immense influence, post-Copernican thinking stayed alive in the nineteenth and twentieth centuries. I shall use this term and the correlated "pre-Copernican thinking" in the following way. Pre-Copernican thinking tends to accept what appears to be object-sided, and thus real, as indeed being *purely* object-sided. It tends to trust our perception that suggests to us that things are as we perceive them.[19] By contrast, post-Copernican thinking systematically considers the possibility that what appears to be purely object-sided (and thus real) may contain genetically subject-sided components. Note that post-Copernican thinking, as I refer to it here, does not dogmatically claim that what appears to be object-sided, and thus real, necessarily contains genetically subject-sided components. It only claims that contrary to all impressions, the apparently purely object-sided, and thus real, *may* contain genetically subject-sided contributions; it is a fundamentally self-critical stance. For those people who have adopted post-Copernican thinking, pre-Copernican thinking in our days appears to be obsolete and dogmatic, "unphilosophical," or even "anti-philosophical" (see, e.g., Rowbottom 2011 against Sankey 2008). The reason is that we know since the advent of Copernicus's theory that apparently purely object-sided real phenomena may contain genetically subject-sided components.

In the nineteenth century, elements of Kant's post-Copernican thinking substantially entered, mainly through the channels of neo-Kantianism, the debates of the emerging historical humanities and social sciences, as well as the non-presentist conception of historiography. These elements stayed there and were partly even radicalized, in the last third of the twentieth century due to a particular reading of Kuhn's philosophy of science. Space limitations prevent me from discussing these developments. Instead, I will immediately jump to the physics of the twentieth century.

---

[19] The German word for perception, *Wahrnehmung*, nicely expresses this property of perceptions. Literally, *wahrnehmen* means something like "take the true" or "take as true." – See also Chirimuuta 2015, pp. 29–30.

1.5 Twentieth-Century Physics

Elements of post-Copernican thinking entered into two of the most important developments of twentieth-century physics, special relativity theory and quantum mechanics. I can only sketch the basic idea.

It is one of the central assumptions of classical physics that it is an objective property of two events to be simultaneous or not. This is also taken for granted by common sense. However, Einstein realized that the simultaneity of two events is not absolute, but relative to an observer. In other words, the simultaneity of two events is not purely object-sided but contains a genetically subject-sided element, namely the state of motion of the observer. The consequence is that one and the same pair of events may be simultaneous for one observer and not simultaneous for another observer. Thus the seemingly purely object-sided simultaneity contains genetically subject-sided components, which are invisible at low relative velocities to the observers.

The role of the observer in quantum theory is even better known, which in the 1920s and 1930s led to fundamental discussions among physicists about the nature of reality. Here is one example. Newton conceived light as a stream of particles; in the nineteenth century, however, the idea prevailed that light is an electromagnetic wave. As became clear in the early twentieth century, light has both properties in irreducible ways, which appears to be a contradiction. It turned out that these irreconcilable properties of light cannot lead to a contradiction because their appearance depends on specific experimental arrangements. The experimental arrangements that make light to show its wave-like character are not compatible with the experimental arrangements that make light to show its particle-like character (see, e.g., Hoyningen-Huene 1994, pp. 241–245). Thus, the presumed fundamental objective characteristic of light, be it to consist of particles or of waves, turned out to be observer dependent. Thus, what seemed to be a purely object-sided property of light (wave or particle) turned out to contain genetically subject-sided elements, namely the specific observational conditions. Again, we see the fundamental ingredient of post-Copernican thought at work.

1.6 Kuhn's Metaphysics

Finally, I can consider Kuhn's metaphysics. The claim of this chapter is that one should understand Kuhn's very insufficiently worked out metaphysics as standing in the genealogy that I outlined in the previous sections.

Kuhn's metaphysics contains the fundamentally post-Copernican element that something that appears to be purely object-sided may also contain genetically subject-sided components. Note that putting Kuhn into the post-Copernican genealogy is, of course, not by itself an argument for its correctness. Understanding the post-Copernican genealogy only opens up a conceptual space that might otherwise be barred: that something objective and real may still contain genetically subject-sided elements that do not diminish its reality status. Familiarity with the post-Copernican genealogy may prevent immediate dismissal of philosophical positions that do look absurd from the viewpoint of common sense and of more or less naive forms of realism. However, any claim that this or that object that appears to be completely object-sided also has genetically subject-sided components must be argued for separately. In no way is post-Copernicanism a carte blanche for antirealism.

From 1979 onward, Kuhn describes his position as "also ... Kantian but ... with categories of the mind which could change with time as the accommodation of language and experience proceeded. A view of that sort need not, I think, make the world less real" (Kuhn 1979, pp. 418–419). This is reminiscent of Kant's description of his position as *simultaneously* transcendentally idealist and empirically realist.[20] Peter Lipton found a nice expression for Kuhn's position: It is "Kant on wheels" (Lipton 2003). Interestingly, this view had been anticipated by Einstein already in 1949 showing that it is not entirely far-fetched for a reflective physicist (see Oberheim 2016, p. 23).

Nevertheless, the parallel between Kuhn and Kant is not total with respect to the objects that contain genetically subject-sided components. Kant's forms of intuition and categories of thought are responsible for the constitution of *what physical things in general are*, that is spatiotemporal things. Thus, Kant's forms of intuition and categories of thought are constitutive of "thinghood."[21] By contrast, Kuhn's view is wider. For him, the constitution of physical things by genetically subject-sided contributions is only a special case. It only applies to the transition of classical to quantum mechanics, where the classical notion of a physical thing is abolished in favor of something like a "quantum object." In most other cases, a revolution does not affect thinghood itself, but only the existence, the qualities, and the relations of *specific things*.

---

[20] For discussion and quotations, see, e.g., Allais 2015, Part Three.
[21] See Heidegger and Gendlin 1985 who particularly stress this aspect.

However, in a fundamental respect Kuhn's spirit is the same as Kant's: What we *correctly* take as objective reality is nevertheless somehow shaped by genetically subject-sided components.[22] As in Kant's case, one can understand Kuhn's position by means of the secondary qualities analogy; for example, in terms of colors (see also Hoyningen-Huene and Oberheim 2009, p. 207). Suppose colors as phenomenal qualities of things are secondary qualities: They are then an amalgam of genetically object-sided and genetically subject-sided components.[23] Clearly, common sense and many sciences take colors unhesitantly for real. For instance, in paleontology the question, what the colors of dinosaurs were, appeared to be a scientifically unanswerable question for decades (all colors we see in images of dinosaurs are completely arbitrary), because these organically based colors are not preserved in the fossil record. However, in 2010 scientists found a way to reconstruct the colors of some dinosaur species, assuming the reality of those colors and their underlying molecular basis as a matter of course (Zhang et al. 2010). Since the seventeenth century, colors were identified as secondary qualities and their underlying primary qualities were investigated. Thus, the reality status of colors has been somehow undermined by the existence of underlying primary qualities (that are supposed to be purely object-sided) without, however, making the secondary qualities completely unreal. The downgrading of the reality status of secondary qualities is the effect of seeing them in the light of underlying primary qualities; without this perspective they successfully pass the reality test by being causally efficacious (in the same way as the apparent motions are).

This analysis enables us to put the metaphysics that Kuhn attempted to articulate into the tradition of post-Copernican thinking, thereby making it at least intelligible. Imagine that all observable and theoretical properties of physical things and processes are secondary qualities like colors, but as robust as the apparent motions of celestial bodies.[24] Now assume further that we have no access whatsoever to their purely object-sided components, that is, to the corresponding primary qualities. Now remember the

---

[22] This is what realists get notoriously wrong. For instance, Bird 2003, p. 691: "Most commentators take Kuhn's term 'world' not to mean the world of things but a world of appearances or of subjective connections (e.g., Hoyningen-Huene 1993)." I did absolutely *not* mean to oppose "appearances" to "things."

[23] This seems to be standard view in color science, but less so in color philosophy; see, e.g., Giere 2006, chapter 2; Chirimuuta 2015.

[24] Colors vary with individual observers and with observing conditions. This reduced robustness has fed the suspicion that they are not fully "objective," this is, purely object-sided; see, e.g., Chirimuuta 2015, p. 6.

treatment of apparent motions and secondary qualities in the history of Western thought before the discovery of their nature as apparent or secondary, respectively. As a matter of course, these motions and properties were taken for real both in everyday life and in science and philosophy. The main criterion for their reality very probably was their causal efficacy, a criterion that was then as persuasive as it is today. Under these circumstances, we would take all observable and theoretical properties of physical things and processes simply for real, despite their true nature as secondary qualities, which in the given scenario are inaccessible to us. Even speculations about possible underlying primary qualities would be completely unhelpful. If we have in principle no access to these primary qualities, their existence would not change our epistemic situation in the least, in comparison to the situation in which the secondary properties were, in fact, primary.

I am now suggesting that already in 1962, Kuhn had this scenario dimly in mind and that he tried to develop it during the rest of his life. Here is a central quote of SSR that is hardly intelligible by itself but is consistent with the given scenario: "In so far as [the scientists'] only recourse to [the world of their research engagement] is through what they see and do, we may want to say that after a revolution scientists are responding to a different world" (SSR-2, 111).

The "only recourse" refers, in the given scenario, to the empirical world containing only secondary qualities, without any possible recourse to the underlying primary qualities. Under these circumstances, "we may want to" speak about world change due to a revolution because there is no world accessible to us that is the same before and after the revolution.

1.7 Conclusion

In this chapter, I have tried to put the apparently strange metaphysics that Kuhn tried to develop from at least 1962 on, in a larger historical context hoping to make Kuhn's attempted metaphysics more intelligible. Some commentators tried to circumvent the difficulties of Kuhn's strange worldchange talk by downgrading it from metaphysical to psychological and metaphorical. For example, noted philosopher of science Alexander Bird writes: "In summary, a change in paradigm can bring with it a range of important *psychological* changes that have cognitive (and emotional) consequences [...]. It is these *psychological* changes that Kuhn is referring to with the *metaphor* of 'world-change'." (Bird 2012, 869, my ital.) And "'World-change' focuses on the *psychological* consequences of a scientific revolution." (Bird 2012, 871, my ital.) Kuhn, however, stated: "I see no alternative to taking literally my repeated locution that the world changes with the lexicon." (Kuhn 1984, 120)

I hope to have shown that in the given genealogy, serious scientists and philosophers did not identify all changes of genetically subject-sided contributions to phenomena as necessarily psychological. On the contrary, it is the defining characteristic of the post-Copernican development that such changes may concern the very subject matter of the natural sciences, physical reality. Kuhn belongs to this tradition by using the fundamental element of post-Copernican thought and trying to make better sense of the history of science.

Acknowledgment

I would like to thank Eric Oberheim for very useful comments on an earlier version of this chapter.

CHAPTER 2

Kuhn's Kantian Dimensions

Lydia Patton

2.1 Introduction: Two Questions

Two questions need to be considered when assessing the Kantian dimensions of Kuhn's thought. First, was Kuhn himself a Kantian of some sort?[1] There are Neo-Kantian traditions, French and German, from which Kuhn could have drawn, and there are more complicated possible influences. In order to assess Kuhn's "Kantian" influences, then, we need to figure out whether Kant, Kantians, or Neo-Kantians have had an influence on his work, directly or indirectly. A second question is whether Kuhn had an influence on the Kantian and Neo-Kantian thought that came after him. Here, rather than looking at sources and influences, we consider a reception tradition.

While both questions are significant to the current project, it is crucial to distinguish them from each other. In particular, as I will argue, making clear distinctions here will allow us to see how recent assessments of Kuhn can be correct, within their own domain, but incorrect when applied outside it (Kant scholars will recognize this move from the Antinomies!).

More and more scholars are working on Kuhn's influences, scholarly development, and reception. For instance, we might look at the competing assessments of Pihlström and Siitonen (2005) and Mayoral de Lucas (2009). Pihlström and Siitonen argue that

> [e]ven though the logical empiricists dispensed with Kantian synthetic a priori judgments, they did maintain a crucial Kantian doctrine, viz., a distinction between the (transcendental) level of establishing norms for empirical inquiry and the (empirical) level of norm-governed inquiry itself. Even though Thomas

---

[1] According to his own records, Kuhn did not read the *Critique of Pure Reason* in toto, at least before March 1949 (Galison 2016, 50). But in 1940 he did take a course that covered Kant, who came as a "revelation."

> Kuhn's theory of scientific revolutions is often taken to be diametrically opposed to the received view of science inherited from logical empiricism, a version of this basically Kantian distinction is preserved in Kuhn's thought. In this respect, as Friedman has argued, Kuhn is closer to Carnap's theory of linguistic frameworks than, say, W.V. Quine's holistic naturalism. Kuhn, indeed, might be described as a "new Kant" in post-empiricist philosophy of science. (2005, 81)

Juxtapose that with Juan Mayoral de Lucas's recent evaluation:

> [T]here seem to be good prospects of a reinterpretation of Kuhn's philosophical thought in terms of an intermittent conversation with Lewis (among other thinkers). The main consequence would be a change of the tradition to which we usually ascribe Kuhn's thought – from French neo-Kantianism to the American (or rather Lewisian) pragmatism and its criticism of Kant. This paper suggests to focus further examinations (and criticisms) of Kuhn's philosophical legacy from this point of view – that is, by assessing his place in the development of pragmatism and the influence of Lewis's pragmatist approach on him. (2009, 183)

Both of these are correct. Friedman, Pihlström, and Siitonen are correct that there is much for the Kantian to mine in Kuhn. Kuhn's distinction between frameworks, or paradigms, and empirical, informal actions of inquiry and investigation is consistent with a Kantian picture, as is his view that judgments about scientific objects (ontology), and even the "world" in which scientists work, can be affected by higher-level decisions about paradigms, axioms, and principles. However, to argue that such a Kantian framework can be found in Kuhn is not to prove that it was Kuhn's own view or that Kuhn acquired the view from Kant or the Neo-Kantians.

Mayoral de Lucas is correct, as well, to argue that Kuhn was influenced by Clarence Irving Lewis's pragmatism and by the work of others at Harvard University at the time. In particular, Wray (2016) argues for the influence of James Bryan Conant, and Mladenović (2007) for the influence of Wittgenstein on Kuhn. That influence is interestingly complicated by Kuhn's interest in other, related figures, including his teachers and associates at Harvard. I will argue, in particular, for consideration of his philosophy teacher Raphael Demos and for authors whom Kuhn read carefully in the 1940s, including Susanne Langer and Philipp Frank, who had emigrated to Harvard in the late 1930s.

The chapter that follows will examine each of the above questions in turn. We will look, first, at Kuhn's training and at his influences and will

Kuhn's Kantian Dimensions

investigate to what extent we can find a Kantian influence on Kuhn's work. Then we will examine how the reception of both Kant and Kuhn in the nineteenth and twentieth centuries has affected scholarly evaluations of Kuhn's Kantian aspects. Finally we will analyze themes that emerge from the discussion in the first two.

2.2 The First Question: Kantian Influences on Kuhn?

A raft of collections on Kuhn has sailed in recently, because of the fiftieth anniversary of the publication of SSR in 1962 (e.g., Kindi and Arabatzis 2012, Devlin and Bokulich 2015, Richards and Daston 2016). Possibly as a result, there is a steep increase in rigorous work on Kuhn's origins and development (Mayoral de Lucas 2009 and especially 2017, an earlier work is Andresen 1999), and on the study of his less well-known texts, including the Thalheimer Lectures (Melogno 2019), his later book BBT (Timmins 2019), and the influence of chemistry on Kuhn's development (Wray 2019, Chang 2012a).

All this work has shed significant light on Kuhn's development: in physics and chemistry, in philosophy, in history, and in his conceptions of psychology and social science more generally. The breadth of the new scholarship emphasizes how very much there is to Kuhn, which can present a challenge. Kuhn began his career as a physicist. Still, he took philosophy courses early on. But his reading ranged from Gestalt psychology, to Jean Piaget, to Alexandre Koyré, to Susanne Langer, to John Dewey. Sorting out the influences on Kuhn is no easy task. In this section, I will limit myself to explaining what we can infer, and what we ought not to infer, about Kuhn's relationship to Kant from recent work on his origins and development as a scholar.

2.2.1 What Is Kantian about Kuhn?

Before we begin, I will provide a sketch of ways that we might find connections between Kuhn and Kant and will list a number of links between them that I think are not justified.[2]

Possibly "Kantian" or Neo-Kantian aspects of Kuhn include:

---

[2] The section that follows makes sweeping assertions that do not divide up Kuhn's works by time period. I have found it necessary to simplify things a bit in order for this topic to become manageable, and I recognize that there is much more nuance to be found in the changes to Kuhn's views over time.

- The scientific paradigms of SSR can function as background explanatory frameworks, conditions for making sense of scientific assertions about objects and phenomena. They are, in this sense, conditions for the understanding and justification of objective knowledge, a position that has been linked directly to the Kantian tradition.
- Similarly, Kuhn argues that there are conditions for knowledge, science, and even experience, in the following limited sense. There are conditions that must be met before we can even make assertions about or have access to the phenomena, and those conditions may involve practical capacities (working with scientific instruments, for instance) or may involve reasoning about scientific laws or proof structures. This position is not necessarily part of Kantian orthodoxy but is quite consistent with the Kantian position.
- Changing one's conceptual, law-governed frameworks has an impact on one's judgments about objects and events and even on one's fundamental ontology. While this view usually is ascribed to thinkers influenced by Kant who were critical of Kant's transcendental idealism (e.g., Hans Reichenbach), it is less foreign to Kant himself than one may think.
- There can be competing frameworks for scientific and epistemological explanation, which cannot be resolved by rational means. While this is often cited as an un-Kantian aspect of Kuhn, it is – of course! – very Kantian.

Elements of Kuhn's view that cannot count as Kantian (at least, not without quite a bit of further explanation) include:

- We can be initiated into – can learn – new paradigms. Kuhn is among those who mention the "context of pedagogy," that is, the context in which scientists learn how to do science, both practically and theoretically. Kuhn's practical approach to learning, and his view about how paradigms go beyond the merely theoretical, is much more pragmatist than it is Kantian (see, e.g., Patton 2017, Rouse 1998).
- Learning, as a way of thinking and practicing, involves being initiated into a community of practice and even involves different standards and requirements for proof and persuasion. As we will see, Kuhn had any number of possible philosophical sources for this view, including Demos, Lewis, and Wittgenstein, not to mention scientific and sociological sources like Fleck, Polanyi, and Toulmin.

At least one aspect of Kuhn is deeply ambiguous. It is sometimes cited as a kind of Kantianism, and Kuhn seems to have acquired it, as we will see, partly from the Neo-Kantian tradition. But it is not really a Kantian view:

- There is no "real" form for any given matter or event. Instead, we structure reality with concepts, serial form, and symbolization (Ernst Cassirer, Susanne Langer, Jean Piaget). In my view, this position comes more from nineteenth-century psychology, including Piaget but especially Helmholtz, and associated rejections of "direct realism" about perception, than it does from Kantian philosophy. In Langer's case, this rejection of direct realism is linked to her connections with Sheffer, Lewis, and Wittgenstein, whom Mayoral (2009) cites as influences on Kuhn as well.

Finally, there is a Kuhnian element that is in no way Kantian and may even be anti-Kantian. I mention it because, while it is missing in Kant, it resembles views found in the Neo-Kantian tradition and is a welcome complement to Kantian views:

- Historical explanation is not a matter of retracing the unfolding of rational principles in time. Rather, history – including the history of science – itself requires a kind of initiation and learning: for instance, about the standards for practice, proof, inference, and reasoning that held sway for a given community. We find this view especially in ET.

2.2.2 *Paradigms as Explanatory Frameworks*

In the 1940s, Kuhn studied physics and philosophy at Harvard, taking one year of philosophy as an undergraduate, including an important course with Raphael Demos (Sigurðsson 1990, 19). In that course, Kuhn said, he studied Descartes, Spinoza, Hume, and Kant.

> Spinoza didn't hit me very hard, Descartes and Hume were both in the current, I could understand them easily; Kant was a revelation ... I gave a presentation on Kant and the notion of preconditions for knowledge. Things that had to be the case because you wouldn't be able to know things otherwise .... It just knocked me over, that notion, and you can see why that's an important story. (Kuhn 2000b, 264)

During the war, he read Percy Bridgman, Philipp Frank, and Bertrand Russell. In the late 40s, Kuhn began to teach in James Conant's General

Education science curriculum at Harvard, which had a profound influence on his development (Nye 2012, Wray 2016).³ Following Conant's suggestion that Kuhn "go find out" about the history of mechanics, Kuhn began reading the historian Alexandre Koyré's *Études galiléenes* (1939), as well as Aristotle, which led to his famous realization that Aristotle's physics is intelligible on its own terms (Sigurðsson 1990, 20).

In 1951, Kuhn gave the Lowell Lectures, which show the deep influence of Koyré (Galison 2016, 60 and n38).⁴

One clear case for an influence of Kant on Kuhn is Kuhn's view that paradigms provide rules that determine how we use concepts and that these rules and concepts are constitutive of objects of experience and knowledge. Kant argues that "[a]ll intuitions, as sensible, rest on affections, concepts therefore on functions. By a function, however, I understand the unity of the action of ordering different representations under a common one" (1781–87/1998, A68 / B93). Kantian concepts and judgments are functions for the unification of representations. Those functions are on the side of the subject and so, in Kant's sense, are a priori: Their unifying actions are independent of experience.

Kuhn's account of how paradigms⁵ allow scientists to become aware of significant phenomena, to solve scientific problems, and to make judgments about objects is an obvious candidate for a Kantian influence. And we might see Kuhn's claim that paradigms provide rules for scientists to follow in solving problems as a key justification for attributing such an influence to Kant (see, e.g., Richardson 2002). Paradigms are not logical functions of unity like Kantian concepts. But they do provide ways of approaching the phenomena: Kuhnian paradigms give a scientist "access to the (region of the) phenomenal world relevant to the work of his or her community" (Hoyningen-Huene 1993, 187). That access is mediated by ways of seeing and recognizing objects in terms of a background

---

³ "At this time the major topics in the physical sciences, and the ones mentioned by Kuhn in SSR, included mechanics and astronomy from Aristotle to Newton, the chemistries of phlogiston and oxygen, Daltonian atomism, eighteenth-century electrical researches, gas theories from Boyle to Maxwell, theories of heat, light, and radiation from Newton through Maxwell, X-rays, quantum mechanics, electromagnetism, and relativity theory" (Nye 2012, 558).

⁴ In 1957, Kuhn published CR, which also shows the influence of Koyré. In the same year, two volumes of Conant's *Harvard Case Histories in Experimental Science* were published. Kuhn's use of the "Copernican Revolution" as a case study obviously has Kantian overtones, but Kuhn's position in the book is much more empiricist. In fact, later, Kuhn is critical of Koyré's excessively a priori take on Galileo (see, e.g., Pinto de Oliveira 2012.)

⁵ As Margaret Masterman (1970) has detailed, Kuhn uses the word "paradigm" in more than twenty different ways in SSR.

framework, the paradigm: which we could view as a kind of naturalized Kantian a priori (see § 2 for details of this notion).

There are a number of obstacles in the way of such a reading, at least in terms of assessing influence on Kuhn. The first is that it is unclear how much of the first *Critique* Kuhn had read, at least before SSR. He took an undergraduate course at Harvard that covered Kant but marked in his journal that he had not read the first *Critique* all the way through, nine years later (Galison 2016, 51).

The second obstacle is that, in §§ 3–6 of SSR, and later, Kuhn qualified the role of rules in the progress of science, even under a paradigm.[6] Kuhnian paradigms should be understood in the context of scientific practice.[7] Even in 1962, Kuhn does not see a paradigm as merely a background theoretical framework. Paradigms are based on model scientific achievements that become organizing norms for scientific communities.[8]

Kuhn does not allow for a Kantian account of perceptual experience, according to which we use concepts as schemata, or rules, for the synthesis and unification of perceptions and representations into experience and judgment. Kuhn was reading Jean Piaget[9] in the 1940s (Galison 2016, 51). Kuhn read Piaget from very early on. In describing Kuhn's "psychologically inflected Neo-Kantianism," Galison notes that for Kuhn, the process of moving from sensations to events and phenomena is

> a continuing one, forged by the encounter of the psychological with the logical and scientific apparatus that brings us the physical visible world. Throughout, Kuhn followed Piaget in presenting a bilayer analysis: on the one side there was the physical world and, on the other, its not-always matched representation in the verbal-psychological. Indicating the relative autonomy of the psychological world, Kuhn pointed out, tentatively, there could be other such "worlds" including "the aesthetic & ethical". (Galison 2016, 53)

---

[6] In SSR, "paradigms can guide science even without rules (§ 3).... [How?] This question is answered with recourse to Ludwig Wittgenstein's theory of concept use via family resemblance (§§ 4–6)" (Hoyningen-Huene 2015, 186).
[7] Including Andersen (2000), Brorson and Andersen (2001), Patton (2017), Rouse (1998, 2013).
[8] In a later interview reporting on a conversation with Masterman, who advocated that paradigms are entirely social, Kuhn even says, "I can't make [what she said] work quite but it's very deeply to the point: a paradigm is what you use when the theory isn't there" (Kuhn 2000b, 300).
[9] "Piaget he did read ... focusing on the psychologist's 1946 *Judgment and Reasoning in the Child* as well as 'Notions de vitesse et de movement chez l'enfant'" (Galison 2016, 51).

Galison describes this as "psychologically inflected Kantianism (sharp division between world and representation)" (Galison 2016, 53). And the account does rely on the engagement between the mind and the "logical and scientific apparatus that brings us the physical visible world" (ibid.), which may sound something like a Kantian a priori or one of its later variants.

The process Kuhn is describing here, of learning to use tools, logical and physical, to bring about a world of phenomena that are available to scientific research, can be read in a Kantian way. But it can be read in other ways. In Piaget's work, it means that children learn about concepts, like motion and speed, not just by developing their psychological and logical capacities, but by being initiated into a set of practices that must mature over time. Piaget did not agree with Kant's implicit characterization of mental "faculties" or "capacities" as operating independently of experience and, in particular, independently of communication with others. In my view, Kuhn's account of paradigms as frameworks for constituting knowledge fits much more naturally within Piaget's perspective than it does within a Kantian account. As Kuhn remarks in a later address, Piaget's

> perceptive investigations of such subjects as the child's conception of space, of time, of motion, or of the world itself have repeatedly disclosed striking parallels to the conceptions held by adult scientists of an earlier [historical period][10] ... Almost twenty years ago I first discovered ... both the intellectual interest of the history of science and the psychological studies of Jean Piaget. Ever since that time the two have interacted closely in my mind and in my work. Part of what I know about how to ask questions of dead scientists has been learned by examining Piaget's interrogations of living children. I vividly remember how that influence [of Jean Piaget] figures in my first meeting with Alexandre Koyré, the man who, more than any other historian, has been my maître. I said to him that it was Piaget's children from whom I had learned to understand Aristotle's physics. His response – that it was Aristotle's physics that had taught him to understand Piaget's children – only confirmed my impression of the importance of what I had learned. (ET, 21–22)

Kuhn here ascribes to Piaget his famous hermeneutical moment, of learning to understand how Aristotle's physics works on its own terms. While Piaget certainly had sympathies with Kant, his influence is not Kantian in this context.

---

[10] ET translates this as "age," but I find that too easily confused with a person's age.

Having said that, it is also true that Kuhn rejected any version of direct realism that would have it that we derive direct knowledge of the real concepts, patterns, and formal properties of objects and events from experience. In that sense, Kuhn's work is very much consonant with the Neo-Kantian philosophers, including Hermann Cohen and Ernst Cassirer, who rejected the "copy" theory that concepts are direct copies of sensations. Thus, there is reason for ascribing something like "Neo-Kantianism" to Kuhn. Richardson (2003) and Patton (2004), among others, detail the resistance to the "copy theory" of concepts, according to which our concepts of things are direct copies of those things' properties. Richardson also details the extent to which the Neo-Kantian Marburg School sees the development of the mathematized science of nature as a necessary a priori condition of empirical knowledge.[11]

And we find an influence on Kuhn in the Harvard context who was directly responding to Cassirer: Susanne Langer. Langer also engaged critically with Wittgenstein's philosophy as early as 1926, including connections with others Kuhn was reading at the time, like Sheffer and Russell (Felappi 2017, 39–40). These early articles were worked into material found in Langer's *Philosophy in a New Key*, which Kuhn read in toto in 1949 (Galison 2016, 51). There, Langer remarks, "The study of symbol and meaning is a starting-point of philosophy, not a derivative from Cartesian, Humean, or Kantian premises" (Langer 1942/1951, viii). Cassirer and Langer focus their attention on how meaning is symbolized in the initial, nonconceptual encounter between humans and the world, through metaphor, myth, and natural language, for instance. The "expressive" element of human experience does not require prior conceptualization or symbolization: it is what is found in myth, for instance.

Langer presents an account of symbolization according to which conceptions of a thing are constructed by a process of abstraction, and "our ability to talk about" the same thing is based on recognition of the pattern in common between several abstractions (1942/1951, 68–70). Langer's remarks are significant for the following reason. In 1961, Kuhn circulated a draft of SSR now known as "Proto-Structure" (see Galison 2016; and Hoyningen-Huene 2015). In Proto-Structure, there was no mention of Wittgenstein or of concept use via family resemblance. According to Hoyningen-Huene, in the 1962 edition of SSR, "Rules of normal science,

[11] Richardson 2003, 61–62 and passim: "[F]or Cassirer, space, time, magnitude and functional dependence of magnitudes are a priori not because we have knowledge of space, time, number, and function independently of experience but because objective experience is first possible through mathematized sciences of nature" (Richardson 2003, 62).

if existent, are derived from paradigms (§ 2). However, paradigms can guide science even without rules (§ 3) ... what does this statement mean? This question is answered with recourse to Ludwig Wittgenstein's theory of concept use via family resemblance (§§ 4–6)" (Hoyningen-Huene 2015, 186). Kuhn gave the MS of Proto-Structure to Cavell, who, Hoyningen-Huene suggests, may have recommended the Wittgensteinian move.[12] Now, "Kuhn says in a taped interview ... that initially he was not aware of the parallel of his paradigms to Wittgenstein" (Hoyningen-Huene 2015, 188; see Kuhn 2000b, 299). But Langer's account of concepts is not very far off, and she mentions symbols as possessing a "visual Gestalt" (Langer 1942/1951, 72).

In any case, Kuhn was a lifelong opponent of the view that science consists of a set of a priori rules that come pre-set, already valid for application to experience, and that do not require initiation or training into ways of seeing, perceiving, and reasoning. Marcum (2012a) details how

> Kuhn contrasts his approach to the approach of traditional philosophers of science who presume that the "process of attaching symbolic and verbal expressions to nature is entirely governed by definitions and rules, explicit or implicit" (1967 TSK Archives, card 16). Although Kuhn acknowledges that definitions and rules do function in science, they cannot account completely for how scientists go about setting up problems and solving them. Often, scientists exhibit unanimity vis-à-vis practice even though they may not entirely agree on the meaning of theoretical expressions. According to Kuhn, solved problems function by allowing scientists to see their way to solving new, unsolved problems. In other words, scientists recognize a similarity relationship between the solved and unsolved problems. Important in this process, Kuhn stresses, is "the learned perception of likeness or similarity [that] is prior to and does not imply the existence of a set of criterion [sic] which would provide a basis for the judgment of likeness" (1967, card 18). (Marcum 2012a, 50–51)

Despite Kuhn's famous emphasis on the priority of paradigms, this final sentence is a clear reflection of his debt to Piaget. And, in fact, Kuhn concludes his 1969 lecture, "Second Thoughts on Paradigms," "by comparing a science student's problem-solving ability to that of a child learning to assemble a puzzle, thereby emphasizing the logical and psychological priority of similarity perception" (Marcum 2012a, 54, see Kuhn 1969/1974). The appeal to a case study involving children is a clear reference to Piaget.

---

[12] Hoyningen-Huene 2015, 188. See Heilbron 1998, Stone 2018 for Cavell and Kuhn.

Kuhn's admiration for Kant is tempered, throughout his career, by his desire to understand the processes of learning, of initiation into a scientific community, of experimentation using instruments, and of persuasion (see § 1.5).

2.2.3 Pragmatic and Conventional: A Priori Reasoning

> Spinoza didn't hit me very hard, Descartes and Hume were both in the current, I could understand them easily; Kant was a revelation ... I gave a presentation on Kant and the notion of preconditions for knowledge. Things that had to be the case because you wouldn't be able to know things otherwise .... It just knocked me over, that notion, and you can see why that's an important story ... I go round explaining my own position saying I am a Kantian with moveable categories. It's got what is no longer quite a Kantian a priori ... I do talk about the synthetic a priori.
> (Kuhn 2000b, 264)

Kuhn does not allow for the possibility of a universal rational framework that proves anything like Kant's results about the status of rational agents or the necessity of the results of Euclidean geometry or Newtonian physics.

However, in the Harvard environment of the 1940s and 1950s, there was no lack of figures who were rethinking the Kantian a priori. The American Pragmatist C. I. Lewis adumbrated, at the time, a "pragmatic a priori" that certainly has deep similarities to Kuhn's account expressed earlier: that the a priori consists of changeable orientations to experience, which one learns. Mayoral (2009) rightly cites Lewis as a possible early source of Kuhn's "moveable categories."

Another, perhaps more surprising source for Kuhn's development was the logical empiricist Philipp Frank. Frank had emigrated to the United States, where he and Hania Frank became refugees following the invasion of Prague (see Holton 2006). He was a physicist and a philosopher (for his philosophical positions see Mormann 2017, Uebel 2012). Frank remembers, of the first Vienna Circle:

> Our whole group understood and fully agreed that the human mind is partly responsible for the content of scientific propositions and theories ... We admitted that the gap between the descriptions of facts and the general principles of science was not fully bridged by Mach, but we could not agree with Kant, who built this bridge by forms or patterns of experience that could not change with the advance of science.
> (Frank 1949, 7–8)

As Uebel notes, it was this context that led Frank's Vienna Circle colleague, Hans Reichenbach, to distinguish between "axioms of coordination" and "axioms of connection" (Uebel 2012, 8). It is well known that Kuhn read Frank early on; they were at Harvard together in the 1940s; and they were wrestling with some of the same questions. Frank was even involved in Conant's General Education curriculum. It is entirely possible that Frank's attempt to "bring about a rapprochement between the logical empiricism of the Vienna Circle in exile and American pragmatism" had an effect on Kuhn (Mormann 2017, 56). However, if so, this would have led Kuhn away from a Kantian a priori and toward something closer to Poincaré's conventionalism.

2.2.4 Paradigms and Persuasion

This section will be quite speculative. It will investigate "a provocative but unfortunately neglected essay by the philosopher Raphael Demos entitled 'On Persuasion' (1936)," which anticipates a number of Kuhnian themes. Demos was Kuhn's first philosophy professor and taught a class that Kuhn says was very important for his development. In the essay, Demos argues

> against sheerly relativistic ("ethnocentric") notions that theory "creates" its facts simpliciter, and thus circularly confirms itself in a prison house not of language but of cultural practices, Demos argues both that divergent theories or worldviews may equally explain the "facts" and that "to explain facts is not enough, a theory must be true" (226) ... such truth is not an easy matter of superior arguments, since the very ideas of argument and reason are contested. (Jost and Hyde 1997, 10)

Page 228 of Demos's essay contains, basically, Kuhn's argument about how paradigms are overthrown:

> In the end ... the theory is imperceptibly worn away by the cumulative force of minute considerations, so much so that it is impossible to put one's finger on the exact factor which led to the abandonment of the theory. There is no distinguishable straw which is the last straw. For the most part, a general pattern resists death by allowing itself to be modified; patterns grow, and even when they are seemingly discarded, they are taken up into another and wider pattern ... the victorious strength of the new elements proceeds from the fact that they adumbrate a new and alternative pattern more satisfactory than the one in use. Thus, ultimately, it is a conflict between patterns. Persuasion is not a mechanical process, but a living growth in which elements are gradually assimilated, and ultimately modify those very tissues which assimilate them;

and like all growth, persuasion is unconscious in its greater part. (Demos 1936, 228)

As Jost and Hyde (1997) note, Demos's account has similarities to Wittgenstein's view from *On Certainty* that at the bottom of fundamental disagreements, one is left with only persuasion. In SSR, Kuhn says something quite similar about paradigm choice: It "proves to be a choice between incompatible modes of community life. Because it has that character, the choice is not and cannot be determined merely by the evaluative procedures characteristic of normal science" (SSR-1, 94).

The conclusion of this section, then, is that there are severe limitations to a Kantian reading of Kuhn himself. While he mentioned Kant as an inspiration, Kuhn's emphasis on learning; on activities of symbolization; on paradigms as practical, not just theoretical; and on the social and community aspects of scientific research as constitutive of scientific reasoning are all outside the Kantian perspective. They are not, however, therefore anti-Kantian or entirely inconsistent with Kantian reasoning. In the section following, we will see how Kantian philosophers, and philosophers of science, have grappled with the Kantian themes in Kuhn's work.

2.3 The Second Question: Kuhn's Reception in Kant Scholarship and in the Philosophy of Science

In order to understand how Kuhn was received in Kant scholarship of the 1980s and 1990s, one should know the main trends and concerns of that period. The following sections do not constitute a complete historical or sociological analysis of that time, but gather some of the main themes and questions, and show how they may have influenced Kuhn's reception, including how Kuhn may be presented to contemporary researchers.

2.3.1 Kant Scholarship

Debates between Peter Strawson and Jonathan Bennett, on the one hand, and Paul Guyer and Henry Allison, on the other, defined much of the Anglophone study of Kantian epistemology in the 1980s and 1990s. Strawson and Bennett delved into the question of how much Kantian thinking could be retrieved if we also needed to maintain the progress in philosophy they saw as emerging from the analytic philosophy tradition, especially the work of Frege and Russell, as well as Quine and Putnam. In *The Bounds of Sense*, Strawson argued that the more empiricist elements of Kant – Kant's restriction of knowledge to objects of possible experience, and his rejection of a psychological basis for epistemological justification – were tenable. Strawson accepted something like Kant's empirical realism but rebuffed transcendental idealism: In particular, he targeted Kant's arguments that space, time, and the categories are purely subjective, independent of experience, and a priori.

With *Kant's Transcendental Idealism: An Interpretation and Defense*, Allison responded to Strawson, especially, that transcendental idealism was defensible even on Strawson's own terms, and that it was supported by further arguments, as well. Allison argued that Kant saw space, time, and the categories as "epistemic conditions" for knowledge. Epistemic conditions are hardly "transcendental" in any mysterious sense: They are simply tools that we use to analyze the representations we are confronted with in experience to work up those representations into knowledge. Kant's position comes down to the claim that the tools are not themselves drawn from experience but that any possible knowledge must nonetheless be restricted to objects of experience.

The Semantic Tradition from Kant to Carnap, a posthumously published history and analysis by J. Alberto Coffa, introduced a generation of scholars to the perceived clash between Kantianism and Frege's and Russell's thought, on the one hand, and the semantic tradition of Carnap and Quine, on the other.

The interests of Allison, Strawson, and Coffa converged on a set of key problems: objectivity versus subjectivity; the status of Kantian space and time; the status of Kant's categories (a priori concepts); and the universality and necessity Kant purportedly claimed for a priori knowledge gained through mathematical natural science. It is a somewhat uneasy business to try to assimilate Kuhn into the scholarly framework of late twentieth-century Kantianism. The questions mentioned just now did not markedly interest Kuhn himself. And, insofar as they did, Kuhn was more persuaded by Piaget's account of these phenomena. Moreover, there are other questions in Kuhn that should be of more interest to Kant scholars: the extent, for instance, to which the Antinomies can be viewed as incommensurable paradigms.

2.3.2 The Dynamics of Reason

Around the same time as the debates between Allison, Strawson, Guyer, and Bennett, researchers concerned with the history of "scientific philosophy," and with the history of philosophy of science, began to reconsider the relationships between logical positivism, logical empiricism, Neo-Kantianism, and pragmatism. Among these were the students and intellectual descendants of Hans Reichenbach, including Wesley Salmon, Clark Glymour, Michael Friedman, Alan Richardson, and Alison Laywine.

Friedman began his career with *The Foundations of Spacetime Theories* and later contributed *Kant and the Exact Sciences* and *Reconsidering Logical Positivism*. His work is marked by a deep engagement with logical positivism, which is also true of Salmon and Glymour, but Friedman inaugurated a return, in this tradition, to the close consideration of Kantian and Neo-Kantian philosophy. True to his antecedents, Friedman has not restricted himself to the consideration of what might be considered "internal" questions of Kant scholarship, such as the ones raised earlier in the Allison-Guyer-Strawson debates of the 1980s and 1990s.

Instead, Friedman has focused on the question of scientific change and on the linked accounts of space, time, and rationality we find in Kant, Kuhn, Cassirer, Carnap, and in scientific philosophy generally. Here, he focuses on the "revolutions" in science, including the transformations to the ideas of space, time, and space-time in nineteenth- and twentieth-century physics, and the changes to the notions of causality and ontology, in the quantum theory and elsewhere. Friedman has said, "In the current state of the sciences ... we no longer believe that Kant's specific examples of synthetic a priori knowledge are even true, much less that they are a priori and necessarily true" (2002, 25). Friedman's situation as a researcher of the history and philosophy of physics leads him to target Kant's claim that there can be "a fixed and absolutely universal rationality" that grounds commonality of judgment in epistemology and in the sciences (25). In particular, Friedman's revival of Reichenbach's constitutive a priori and his defense of a dynamic a priori in *The Dynamics of Reason* are inspired, explicitly, by Kuhn's account of changing paradigms.[13] Friedman's "relativized a priori principles constitute what Kuhn calls paradigms: relatively stable sets of rules of the game, as it were, that make possible the problem solving activities of normal science ... In periods of deep conceptual revolution it is precisely these constitutively a priori principles which are then subject to change" (Friedman 2008, 39).

Friedman's account responds to the great revolutions in science in the twentieth century: relativity theory and the quantum theory, and the crisis

---

[13] Mormann 2012 argues that Friedman has underestimated the role of a Lewisian pragmatic a priori in the dynamics of reason and provides a synthesis of the two.

that this provoked in the philosophy of science. Because of this, it is aimed at solving a particular set of problems, problems for which Friedman looked to Kuhn for the answer. Reciprocally, however, Friedman's account has been employed as a way to save Kant from the objection that his universal, necessary a priori was undermined by those same revolutions in science.

2.4 Conclusion: The New Image of Reason in History

Kuhn presents science in a much more messy, historically contingent, and socially charged way than Kant does. Certainly, Kuhnian paradigms provide norms for pursuing science. But Kuhn would not attempt to present, as Kant does, an "ideal history" of reason or a "metaphysical foundations" of the natural sciences. While it is possible to assimilate Kuhn's picture into a Kantian framework, it is crucial to see how the Kuhnian material has to be cut and hemmed first.

A question of interest to both Kuhn and Kant is: what is the status of science, and how do we account for the role of the scientist in its development and justification? Kant developed epistemic, transcendental conditions for knowledge with the purpose of demonstrating the limits of metaphysical reasoning – of providing a critique of pure reason.

For Kuhn, in contrast, the limits of reason are given by the limits of human communication. From Piaget, Kuhn learned that we know only what we can communicate. A child may have a private notion of "motion," or "speed," for instance. But until she can communicate the rational basis for her knowledge, her knowledge cannot be recognized and therefore is not knowledge at all, or even conscious thought.[14] This is a Kuhnian condition of knowledge, but it is not one that Kant cites.

A key question regarding the Kuhnian dynamics of reason, then, is how the capacities for recognition and communication of frameworks, concepts, ideas, and explanations are developed within communities.[15] Such an account requires a much more historically situated and flexible framework. It also requires not just that we develop a relativized a priori, but that

---

[14] Piaget: "[I]f a proposition cannot be expressed, we cannot be conscious of it. When we say that a child can handle a notion before having become conscious of it, what we mean is that there has been gradually built up in the child's mind ... a schema which can be applied ... but which does not yet correspond to a verbal expression. ... The conscious realization of one's own thought is dependent upon its communicability, and this communicability is itself dependent upon social factors, such as the desire to convince" (1928, 29–30).

[15] The account of "SSR as a Weberian Explanatory Model" (Mladenović 2007) is perhaps along these lines.

we develop an account of the development of capacities for reasoning and expression, not just within the individual, but within societies, disciplines, and communities.

The most compelling defense of a Kuhnian approach to the narration of the history of reason, and of scientific and philosophical communities, can be found in the work of Richardson (2012, 2002, 2015). Richardson shares with Kuhn a concern that many histories of science and of philosophy are unhistorical.

> The set of historical facts of the development of science is not, on [Kuhn's] view, primarily a source of evidence for or against a philosophy of science, or for illustration of the workings of old or new philosophical machinery. Rather, history of science as a practice engaged in by historians demands the formation of coherent and explanatory historical narratives and the practices involved in the creation of those narratives themselves demand answers to different sorts of questions than the default philosophical machinery would lead you to ask in the first place. In somewhat different words: For Kuhn, the practice of history – the development of historical understanding itself – stands in complicated but ultimately incompatible relations to the sorts of concerns and the machinery for understanding science posited in the logical empiricist philosophy of science (as he understood it) of the 1950s. (Richardson 2015, 43)

The sorts of questions one should ask, in the history of science and of philosophy, include what Richardson calls "tone-lowering" moves toward a real historical account of human endeavor, rather than an abstract, triumphalist narrative: "[W]e concentrate on persons, organizations, schools ..., on what they were actually doing and what they were actually motivated by; history of philosophy becomes interested in the contingent, the bounded, and human" (2012, 246–247). "Ironically," Richardson continues, such a history will reveal that the target of our study is more interesting than we thought at first. We had thought to "save" the interest of our research by showing it can be derived from a mighty tradition of impressive ancestors. But regarding science, and philosophy, as human endeavors – as attempts to find recognition, to communicate one's results, to find answers to puzzles – reveals the real interest of these pursuits. They are connected vitally to human purposes. "Such tone-lowering gestures in the case of philosophy would, ironically, remind us that in various places and times philosophy has had important and complicated relations with religion, science, art, and other important human activities and endeavors. In some places and some times philosophy has really mattered to human culture" (247). Richardson engages with the heart of Kuhn's SSR: the struggle against the "image of science" as a bloodless collection of results, like lifeless butterflies, each pinned to its classifying card showing its place in a historical timeline. The value of history is that it allows the results and achievements to be seen as they really are. A good history shows not only that an achievement was made, but that it was a real event, connected to human purposes and linked with a much more complex situation than it may seem at first.

Richardson's work, and that of scholars working in the "new historiography of science" including Mayoral and Pinto de Oliveira, emphasizes the requirements of the Kuhnian dynamics of reason: that it recognizes the real conditions for the development of rational capacities, the recognition and dissemination of results, and the satisfaction of human and rational purposes, within scientific and philosophical communities. Such a project goes well beyond anything Kant ever imagined. But it is a contribution, one that still largely remains to be made, to Kant's project of characterizing the scope, limits, and essence of human reason and action.

CHAPTER 3

A Public Intellectual and a Private Scholar
On Thomas Kuhn, James B. Conant, and the Place of History and Philosophy of Science in Postwar America

George A. Reisch

3.1 Introduction: Thomas Kuhn's Disengagement from Politics

If one were to ask what potential role history and philosophy of science can play in civic and political affairs, SSR is an appropriate text to examine. Core assumptions and questions about the nature of scientific knowledge, of scientific change, relationships between the sciences, scientific perception, and rationality itself wind through SSR. In some cases, they first took shape in Kuhn's lucid and engaging analysis of scientific revolutions. To ask whether and in what ways philosophy and history of science can meaningfully engage social and political realities, therefore, is, in large part, to ask whether SSR itself has resources to do this. According to Kuhn, I expect, the answer would be a decisive "no."

In 1973, eleven years after SSR was published, Mr. Kenneth Pietrzak of Hartford, Connecticut, asked a question like this. He had read SSR and was told by one of his teachers that Kuhn had written about politics. Mr. Pietrzak was intrigued enough to write to Kuhn and ask whether such an article existed and, if so, where he might find it. Kuhn replied that politics appeared briefly in SSR (presumably he had in mind the first pages of "The Nature and Necessity of Scientific Revolutions," where he lays out the formal similarities between scientific and political revolutions). But he dismissed these pages as "a few scattered remarks." He also downplayed the "bit of information" on politics in SSR's postscript (where he compared SSR to histories of political systems) and then signed off: "Beyond that there is nothing. I cannot imagine what your informant had in mind."[1]

---

[1] Kenneth Pietrzak to Kuhn, 1973; Kuhn to Kenneth Pietrzak, 1973. TSK Archives – MC240.

Having gone through much of Kuhn's correspondence, I have become familiar with the "intricate embroidery" of his personality, as Norwood Russell Hanson once described it.[2] On that basis, Kuhn's reply says more than simply "no, you are mistaken – I haven't written on politics." His reply says that the inquiry's premise is slightly fantastic, even a little ridiculous. Professional scholars like himself, Kuhn seems to think between the lines of his reply, simply do not engage with political questions or issues in any substantive way.

Yet Kuhn did not always believe this. During his childhood in and around New York City, he was familiar with left-leaning teachers and professors who eagerly, sometimes urgently, engaged with political topics and controversies. His parents' close friends included Max Eastman, to take one prominent example, one of the most politically visible philosophers of his generation who edited the socialist magazine *The Masses* in the 1910s. When Kuhn was about ten years old, his parents sent him to the progressive Hessian Hills school, which was dedicated to pacifism, socialism, and other leftist causes. Kuhn thrived, he later recalled, both intellectually and in the "radical" (RSS 256, 257) political atmosphere. The school held a public assembly on Saturday, November 16, 1935, which was devoted to the evils of war. As local newspapers reported the next day, the festivities were led off by a thirteen-year-old Thomas Kuhn who lectured against imperialism and against "the capitalists" who profit "by our national possessions" (Reisch 2019a, 3–4, 9).

At the Taft preparatory school in Connecticut, and as an undergraduate at Harvard in the early 1940s, Kuhn remained politically engaged. He wrote essays on polling, on democracy in America, on the American declarations of war against Germany and Japan, and on his personal struggle to reconcile the pacifism of his high school years with the pro-interventionism he adopted at Harvard. He wrote a poem titled "Civil Liberties" and at his graduation in 1943 he delivered a Phi Beta Kappa address about the "nihilism" and bewilderment of his interwar generation. Kuhn and his peers, he said, were "born into a world not yet recovered from the disappointment and disillusionment of one war" and were now tasked with fighting another.[3]

---

[2] Hanson n.d.
[3] Kuhn 1943. TSK Archives – MC240. Papers on other topics mentioned are contained in TSK Archives–MC240, box 1, folders 2 and 3.

3.2 The Politics of General Education

After Kuhn served as a military radar specialist, evidence of his political engagement becomes thinner. He dedicated himself to completing his physics doctorate and, as a self-described "neurotic" young man, turned inward and underwent Freudian analysis.[4] As his enthusiasm for a career in physics waned, he accepted an offer to teach in the new general-education program founded by Harvard president James Bryant Conant. Teaching and preparing classes within this program, Kuhn became a historian and began to work out the theory of science that would lead to SSR and his other writings.

It would be mistaken, however, to suppose that Kuhn's budding academic career automatically precluded political interests and engagements. For Conant's new program was essentially engaged with issues of the day, both nationally and internationally. To see what motivated Conant to start this program, consider the way things looked to him in 1933 when he first became president of Harvard. Conant was a chemist who deeply admired German chemistry and the educational culture that had made German science possible. So he was especially shocked by Germany's collapse into Nazism. Along with intellectuals of his generation, he struggled to understand this and concluded that Germany failed because, despite its sophisticated scientific and literary culture, it had become susceptible to a powerful ideology that, like a biological virus, had corrupted the German mind.[5]

Conant worried that the same kind of thing could happen in the United States. In newspaper articles and addresses to students, both in the 1930s before America joined the war and in the war's aftermath, he discussed the threat of totalitarianism and the prospect that civic and economic problems could induce the kind of "political mass psychosis" that had beset Germany. Speaking on campus at the beginning of the fall term, just months after the war had ended, he mused about what had happened in Germany:

> Questions like these keep recurring in our minds: How could the Nazi doctrines gain such ascendency among a highly literate and apparently well-educated people? How could that nation breed such rulers and such brutal gangsters and allow them to terrorize the population? One answer to such questions has been given by a learned German professor in a private letter

---

[4] RSS 280; see also Andresen 1999; Forrester 2007.
[5] On Conant's Germanophilia, see Hershberg 1993, 36. On American intellectuals responding to Nazism, see Greif 2015. On Conant's use of the noun "virus" to describe totalitarian ideology, see Conant 1943.

I saw not long ago. The writers philosophizing on the triumph of the Nazi party (which he had done nothing to prevent) explained the situation in these words: "One reason is that the education of the German people was carried out not only in frequently excellent schools but also on the military drilling grounds. Consequence: high mental and intellectual development, great military bravery, yet at the same time lack of civil courage, as many including Bismarck have said before. Lack of civil courage fosters political mass psychosis."

Conant took this diagnosis to be "essentially correct" and explained that an abundance of "civil courage" would be required for the United States to navigate problems he saw on the postwar horizon. "Our present highly industrialized and over-urbanized society provides a medium highly favorable for the development of this disease" (Conant 1946, 224, 226). This favorable medium included as well the nation's immensity, its extremely varied population, and regional differences in economy, culture, and religion that had helped lead to the civil war not yet one hundred years before. These factors stood to be exacerbated by postwar unemployment and the economic turbulence of converting back to a peacetime economy. Early signs of the brewing ideological rivalry between the United States and the Soviet Union were ominous, as well, especially given the educated consensus that the Soviets would soon have their own atomic bomb.

What the nation required, therefore, were educational reforms designed to cultivate and promote civil courage. "To the extent that we see in the downfall of that country [Germany] a failure in education," Conant said in his speech, "we become aware of the responsibility of our schools, colleges, and agencies of adult education." Public education must unify the nation, cultivate widespread understanding of its original democratic and pluralistic values, and help citizens convert from a wartime mentality that prized regimentation and expediency to a mentality of tolerance, pluralism, and democratic process. Only then, Conant reasoned, could the nation handle postwar challenges rationally, deliberately, and democratically – without giving in to "fear, panic, foolish short-sighted action" and other hallmarks of political reaction (Conant 1946, 227).

Two years before this speech, Conant had set these educational reforms in motion. In 1943, in the midst of administrating the development of the atomic bomb, he created a committee at Harvard to study the state of American education and formulate national curricular guidelines to help meet postwar challenges. The committee's final report, colloquially known as the "redbook," appeared in the summer of 1945 (Harvard University 1945). Because the reforms proposed were aimed at the nation, the redbook received national attention. They were also aimed at Harvard's undergraduate education, however, so they were widely discussed and evaluated by faculty, alumni, and students. Conant may well have first learned of Thomas Kuhn when the report was published. For Kuhn, just back on campus after the war, was commissioned to review the *Objectives* report in a Harvard Alumni magazine and, specifically, to offer an undergraduate's perspective (Kuhn 1945).

Science was important to Conant's reforms. The methods and precepts of modern science were vital to a free, democratic society and must therefore be cultivated by any national curriculum. As the redbook put it, "[S]cience is both the outcome and the source of the habit of forming objective, disinterested judgments based upon exact evidence. Such a habit is of particular value in the formation of citizens for a free society" (Harvard University 1945, 50). Accordingly, themes like objectivity, open-mindedness, and aversion to dogma weave through and join the report's discussions about science pedagogy to the character of American culture and history. "Prejudice brings in irrelevancies and logic should keep them out" (Harvard University 1945, 66), the report noted – not *a propos* of scientific research, but rather voting behavior in the American electorate.

Two years after Kuhn reviewed the report, he was asked to join Conant's new program. Conant asked Kuhn to begin preparing for his new career as a historian by reading press proofs for Conant's book, *On Understanding Science* (RSS 275). Originally given as a series of lectures at Yale in 1946, the book explained why educated Americans should know more about science and what a course for future bankers, lawyers, and business leaders might look like. Conant began by alluding to what had been top secret while the *Objectives* report was prepared, namely the atomic bomb whose creation and testing Conant had helped to bring about. Ironically, the new "atomic age" had made Americans more aware of modern science and its fruits, but most were uninformed about how science works and achieves its results. That is why "we need a widespread understanding of science in this country," Conant explained, "for only then can science be assimilated into our secular cultural pattern. When that has been achieved, we shall be one step nearer the goal which we now desire so earnestly, a unified, coherent culture suitable for our American democracy in this new age of machines and experts." (Conant 1947, 3)

The point was not, *per impossibile*, to teach educated Americans intricacies of quantum mechanics or nuclear chemistry. It was to help them better understand scientific method. By examining appropriately structured historical case studies, students would see how progress occurs as theories and experiments mutually inform and revise each other. They would see the creative, scientific imagination at work within the "tactics and strategies" of research; the occasional importance of sheer luck; and how social realities of science and competing schools of thought can inspire as well as mislead researchers, at least in hindsight. This last point, the fallibility of science, is crucial and Conant returned to it a few times. Americans had to see, for example, that research is messy and formula-defying: "The stumbling way in which even the ablest of the early scientists had to fight through thickets of erroneous observations, misleading generalizations, inadequate formulations, and unconscious prejudice is the story which it seems to me needs telling" (Conant 1947, 15).

It needed telling because the case studies he proposed – the discovery of air pressure, of X-rays, and oxygen's role in combustion – illustrate and teach aspects of civil courage, as if laboratories and scientific societies were microcosms or exemplars of the open, dynamic, and courageous society Conant idealized for the nation. These aspects include objectivity, fairness, respect for evidence, toleration of competing views, humility, faith in progress, and healthy skepticism about claims to final truth. Humility, for example, is on display insofar as the case studies highlight the perils of dogmatism and problematize easy answers to who was "right" and who was "wrong" in past scientific disputes. As it was in the famous "Spirit of Liberty" speech that Judge Learned Hand delivered in Central Park a few years before ("The spirit of liberty is the spirit which is not too sure that it is right," Hand insisted), Conant's best scientists pay attention to what they do not know, to "unsolved puzzles" that lay groundwork for future advances (Conant 1947, 91).

Conant's treatment of *progress* involves both the substantive reality of scientific advance, in which Conant steadfastly believed, as well as his recognition that this was a matter of faith that cannot be conclusively demonstrated. Instead Conant invited his readers "to perform an imaginary operation." "Bring back to life" the great scientists of the past, he proposed, and ask *them* to judge whether the science of today had advanced beyond the science they knew. "No one can doubt how Galileo, Newton, Harvey, or the pioneers in anthropology and archaeology would respond" (Conant 1947, 21).[6]

---

[6] This thought experiment reappears throughout Conant's popular writings on science. Useful context for his position is Conant's admiration of Michael Polanyi. See, e.g., Michael Polanyi's essay *Science, Faith, and Society* (Polanyi 1946, 67): "I do not assume that I can force my view on my opponents by argument."

Conant's suspicion of *truth*, finally, points to his devotion to William James, whose famous treatment of truth aligns with Conant's own (Hershberg 1993, 30, 578). Science may be a search for truth, but truths attained are not so much discovered as made from within the scientific process. As James famously put it in his lectures on pragmatism, truth "*happens*" to an idea that is important and useful (James 1981). Not unlike the account put forth by Louis Menand in his *The Metaphysical Club* (Menand 2001) in which the development of American pragmatism and its approaches to truth were conditioned by the American civil war, Conant recognized the concept of truth as a two-edged sword. In science and learning, it inspires constructive intellectual, technological, and cultural progress, but in times of social conflict – and especially in the hands of a demagogue – it can inspire absolutism, fanaticism, and destructive conflict. To be sure, there are conflicts, disagreements, and angry losers in Conant's case studies. But science itself, as his ideal of progress counsels, always succeeds and advances knowledge. Indeed the best possible definition of science, Conant suggested, is that science is an unending process. It "emerges from the other progressive activities of man to the extent that new concepts arise from experiments and observations, and the new concepts in turn lead to further experiments and observations" (Conant 1947, 24).

This was a definition of science for America and the new "scientific age in which we live" (Conant 1947, 25) – an age of wonders but also challenges the nation must surmount without succumbing to violent conflict or "political mass psychosis." That is why instead of prizing "truth" – or, as he pragmatically put it, "the adequacies of our conceptual schemes as to the universe" – Conant's definition prized "the dynamic character of these concepts as interpreted both by professional scientists and laymen. Almost by definition, I would say, science moves ahead" (Conant 1947. 25).

3.3 The Struggle with McCarthyism

Conant was right to worry whether the nation could avoid panic-stricken, reactionary politics. Though he initially resisted, by early 1950, Conant himself joined the growing majority of Americans who believed that the US government had been infiltrated by secretive communists' intent on securing global domination for the Soviet Union.[7] It is not necessary to review Conant's multifaceted career as a cold warrior to see how it contributed to the climate in which Kuhn developed his theories of science.

---

[7] On Conant's conversion to cold warrior in the late 1940s, see Hershberg 1993; Conant 1970, 561.

Perhaps most important is the fact that, despite his national prestige and anticommunist credentials, Conant himself was a target of Senator Joseph McCarthy. As Conant wrote in his autobiography, he "stood high on the list of those whom Senator McCarthy was 'out to get'" (Conant 1970, 564). First, McCarthy pursued Conant as a university administrator too lenient or soft on allegedly suspicious members of his faculty. Conant resigned his presidency in early 1953, shortly before McCarthy and other Washington anticommunists began to scrutinize those faculty members. But in his new diplomatic position as the American High Commissioner in Germany, Conant was arguably more vulnerable for he now worked at the State Department that McCarthy claimed had been overrun by communists. Specifically, Conant was responsible for United States Information Agency (USIA) libraries in Germany when McCarthy claimed to discover that USIA libraries across Europe were providing communist or suspiciously communist books (some 30,000 titles, he claimed). At first Conant remained above the fray and, because he was traveling, avoided Roy Cohn, McCarthy's committee attorney, and his associate David Schine when they toured Europe and visited Conant's commission headquarters. But weeks later, in June 1953, McCarthy attacked Conant at a Senate Appropriations Committee meeting where Conant presented the commission's annual budget. The budget was approved, but this encounter with McCarthy did not go well. In his autobiography Conant reproduced the transcript, presumably to support his evaluation of McCarthy as a "great and malignant opportunist" who wielded Hitler's technique of the "big lie" (Conant 1970, 561, 578).

There is abundant evidence, mostly circumstantial but some direct, that Kuhn paid attention to Conant's career as a cold warrior. First, there is Kuhn's longstanding interest in national and international politics. Second, there is his intense admiration of Conant, one that predates their collaboration, as illustrated by some of Kuhn's editorials for *The Crimson*, the Harvard student newspaper, and Kuhn's positive review of the *Objectives* report in an alumni magazine.[8] Third, and perhaps most important, once their collaboration began, Conant became a professional mentor to Kuhn, and his general-education project set the stage for Kuhn's future career as a professor, a historian, and an expert on teaching science within general education. This particular episode occurred shortly after Conant had left Harvard, but Kuhn's relationship to Conant continued, as

---

[8] Hufbauer 2012, 425–426; Kuhn 1945. On Kuhn's personal admiration for Conant, see also RSS 260, and for a Freudian analysis see Forrester 2007.

evidenced by Conant's foreword to CR, SSR's dedication to Conant, and Kuhn's request to Conant for help getting SSR published – help that in the end was not needed (Reisch 2019a, 271, 283).

The cat-and-mouse game between Conant and McCarthy reinforced at least two lessons for Kuhn. One is that anticommunist politics could reach deep into the academy and was powerful enough to challenge a national leader like Conant. Even before the McCarthy era, Kuhn had observed the headmistress of his progressive school, herself a community leader, lose her job for largely political reasons (Reisch 2019a, 21–24). In 1953, as McCarthy and other congressional investigators scrutinized Harvard by holding hearings on campus, Kuhn knew at least two faculty members who had been targeted: his former physics teacher Wendell Furry and David Hawkins, a visiting colleague in the general education program. Kuhn also knew the cooperating witness Robert Gorham Davis, who had been his undergraduate composition instructor. Davis, a former communist, testified on campus about his membership in a communist cell at Harvard in the late 1930s and named names of about fourteen others who had belonged. Few on campus were likely to have missed the "300 spectators and a barrage of newsreel and television cameras," as the *Crimson* reported, or the local and national press coverage of McCarthy's protracted pursuit of Furry, McCarthy's announcement that the university was a "smelly mess" of communist subversion, and the ensuing public exchange between McCarthy and Conant's successor, Nathan Pusey.[9]

The other lesson was that Kuhn himself was close to one of McCarthy's high-profile targets, perhaps unnervingly close given McCarthy's aim "to get" Conant by discovering communists on Conant's staff or within his organizations. Kuhn was no communist. He was a devout and proud liberal, especially in his youth, as he explained to Gorham Davis years before in one of his undergraduate essays. He was raised and schooled to champion civil and economic rights, to support Roosevelt's New Deal, and to respect "freedom for Communists" to associate and campaign like members of other parties.[10] Still, at the height of the red scare, liberals often came under suspicion for insufficient condemnation of communism or under suspicion as so-called fellow traveling communists who feign liberalism and toleration for the ongoing work of their secret comrades.

In addition, regardless of one's current and stated political beliefs, it was often the politics of one's youth that created trouble. Theodore Kaghan,

---

[9] Crimson 1953; 1953a; 1953b; 1953c. Boston Globe 1953; New York Times 1953.
[10] Kuhn 1941. TSK Archives – MC240.

who worked under Conant in Germany, lost his job after Roy Cohn and David Schine produced information that Kaghan had once written a left-leaning antifascist play. As Cohn's biographer put it, "It was cases like Kaghan – taking middle-aged people and holding them accountable for what they had done in their salad days – which made some people fearful that the young people of the 1950s would be intimidated into sterile conformity" (Von Hoffman 1988, 147). The celebrated cases of faculty members being fired at the University of Washington in 1949 set the tone for the next few years as investigations took place across the nation and during which most scholars chose to lay low and avoid political subjects in their lectures and writings (Schrecker 1986). In 1951 the *New York Times* reported on the basis of a survey it had circulated that "a subtle, creeping paralysis of freedom of thought and speech is attacking college campuses" and that this was largely due to "students' fear of red label." The article described eight kinds of anxiety and inhibition attributable to the climate, one of which was "an unusual amount of seriocomic joking about this or that official investigating committee 'getting you'."[11]

3.4 Politics in the Lowell Lectures

In early 1951, several months before the *New York Times* published this account of the campus climate, Kuhn delivered the Lowell Lectures, a series of eight hour-long talks at the Boston Public Library. The lecture series has a long, illustrious history and usually featured senior scholars and scientists who had made marks in their field. Kuhn was twenty-eight years old, had yet to publish anything in history or philosophy of science, and was understandably nervous. Decades later he recalled that he gave the lectures during his personally difficult years of the late 1940s and early 1950s, saying "I had a dreadful time preparing [the lectures] and I nearly cracked up" (RSS, 289).

To whatever extent Kuhn felt personally vulnerable because of his "radical" past, the lectures show that he understood the perils at hand. In a paragraph elaborating the popular notion that "a scientific discovery must fit the times" or "the times must be ripe," he remarked that "crisis periods" in the history of a field may be sparked or affected by "social forces" or "changes in economic structure" – at which point his typescript reads, "Attention Senator McCarthy."[12] The joke draws upon this anxious

---

[11] Seigel 1951.
[12] This and subsequent quotations from Kuhn's Lowell Lectures come from my forthcoming edition of the lectures. For unedited lecture scripts and outlines, see TSK Archives–MC240, box 3.

mood on American campuses and the well-known Marxist principle that a society's economic base or relations of production shape – and therefore tend to "fit" – the science, philosophy, and arts that develop upon those basic relations.[13] "Attention Senator McCarthy" probably would have generated a laugh, but we do not know if Kuhn made this joke or not. In his archival typescript, the three words are crossed out, almost certainly by Kuhn's hand.

Kuhn also worried that he might be pigeonholed or "tagged" as a left-wing theoretician of science, specifically a theoretician in the mold of Britain's John Desmond Bernal, who urged in his book *The Social Function of Science* (Bernal 1939) that research be organized, guided, or planned to achieve advances in agriculture, medicine, economics, or industrial production that would improve modern life.

To understand this concern, it is necessary to review the treacherous politics of scientific planning in the United States. Advocates of planning were usually socialist, if not Marxist, and took some inspiration from Josef Stalin's five-year plans for modernizing and industrializing the Soviet Union. In philosophy of science, advocates included Otto Neurath and William Malisoff, the founder of the Philosophy of Science Association and founding editor of the journal *Philosophy of Science*. Another prominent advocate was Neurath's American cousin, Waldemar Kaempffert, who became science editor of the *New York Times* in 1931 and used his position to promote Neurath's projects in adult education, his ISOTYPE project, and his encyclopedia of science (Reisch 1994; Reisch 2019b).

Collective organization and planning were prized within Soviet Russia, but critics of planning in the West insisted that it sacrificed essential ideals and methods such as freedom, unfettered curiosity, and laissez-faire administration. Critics also distinguished firmly between pure and applied science and argued that successful applications trickled down from pure, "useless" (Flexner 2017) research and in ways that could not be predicted in advance. Any attempt to command scientists to produce this or that practical result, therefore, put the cart before the horse and was bound to harm scientific institutions and imperil their productivity. In a review of Neurath's new encyclopedia, itself an effort to organize scientists and philosophers to cultivate unity and "build bridges" among the sciences, Kaempffert anticipated and tried to defuse this line of criticism. Comparing individual sciences to individual ships in a flotilla, Kaempffert explained that "the encyclopedia is intended to tell each

[13] See, e.g., Marx 1904, 11–12.

navigator what other ships are doing and what he can learn from their signals, their movements, their errands." Crucially, however, "[t]here will be no admiral to give orders."[14]

Though Kuhn would later write SSR as a contribution to the encyclopedia, it was his Lowell Lectures that intersected with contentious debates over science planning that arose when congress envisioned a national foundation to support scientific research – what became the National Science Foundation. In the mid-1940s, debate was joined by congressional representatives, President Truman, foundation executives, journalists like Kaempffert, and scientists like Conant. They sparred over what the new foundation should look like, how it should operate, and to what extent research supported with public funds should be planned and organized for potential public benefit, at the one extreme, or driven only by the curiosity and talent of scientists, at the other. From his pulpit at the *New York Times*, Kaempffert promoted the collectivist ideals and methods he saw in Neurath's encyclopedia, while Conant and others held firm that the new foundation should support pure science – only – and that any potential benefits to society would follow naturally in due course. In a group letter to President Truman signed by Conant and about forty prominent scientists and administrators, Conant firmly opposed legislation "calling for an over-all director as the supreme authority of all Government-supported scientific research," as the *New York Times* described the letter (Reisch 2019b).

In *On Understanding Science*, Conant noted "the storm signals of present controversy" between purists and planners and respected the arguments on both sides. Given that "echoes of this controversy find their way into the daily press, it is certainly desirable that the student [in general education courses in science] be directed to the writings of the two opposing sides."[15] Other critics of planning, however, were less measured and tolerant. Warren Weaver of the Rockefeller Foundation, a cosigner of the letter to President Truman, continued to attack Kaempffert with methodological arguments overflowing with political passion: "The earnings of science are not to be gained by organizing a super-control which holds guns at the heads of scientists and tells them what to do," Weaver fumed in a letter to the editor. "The earnings of science are gained only by setting the scientists free" (Reisch 2019b).

---

[14] Kaempffert (1938).
[15] Conant 1947, 107. In a detailed endnote, Conant indicated texts by Boris Hessen, J. G. Crowther, Bernal, and the foremost critic of planning in England, Michael Polanyi (Conant 1947, 129–130, n. 12).

A Public Intellectual and a Private Scholar

Having read *On Understanding Science*, Kuhn was aware of this controversy. Its volatility, Conant's public stand against a single-director model for the new foundation, and the increasing disrepute and dangers of political leftism in McCarthy's America begin to explain Kuhn's reaction to the advance publicity for his lectures circulated by the Lowell Institute. Upon seeing a small announcement in the Boston Globe, Kuhn telephoned the institute to demand that they retract the advertisement and its tagline, "What are the problems of scientific research today?" In a letter following up on his telephone call, he attempted to explain his "acute distress." "It may help you to understand my dismay," he wrote, "if I explain that the fascinating topic your copy writer has so clearly stated is one to which I believe no serious and responsible student of science would address himself."[16] Kuhn had earlier provided the institute with a title for his lectures: "The Quest for Physical Theory – Problems in the Methodology of Scientific Research." The institute's tagline, however, failed to mentioned "methodology" as Kuhn's area of expertise. Without this word, his stated expertise in *methodological* problems was represented as expertise in *scientific* problems, and the tagline, to make things worse, suggested that Kuhn's lectures would answer the question it posed. At the dawn of the atomic age, when the public was dazzled and curious about promises made on behalf of nuclear energy and technology, and with Washington having passed the National Science Foundation Act less than a year before, Kuhn not unreasonably saw himself being introduced to the public as an advocate of planning, an expert in "the problems of scientific research today" that will transform and improve the world.[17]

3.5 The Scope and Aim of the Lowell Lectures

The Lowell Institute agreed to pull the advertisement. Weeks later, in his first, introductory lecture, Kuhn set the record straight and introduced his real topic, the methodological study of how knowledge

---

[16] Kuhn (1951a). A month before, describing his current research to one of his Deans, Kuhn made a similar point: Methodological study can help us understand the nature of science, but it "cannot ever venture to prescribe fruitful research procedures to the working scientist" (Kuhn 1951b).

[17] Though he spoke three years later, remarks by Atomic Energy Commission Chairman Lewis Strauss illustrate popular hopes for the new atomic technologies: "It is not too much to expect that our children will enjoy in their homes electrical energy too cheap to meter, – will know of great periodic regional famines in the world only as matters of history, – will travel effortlessly over the seas and under them and through the air with a minimum of danger and at great speeds, – and will experience a lifespan far longer than ours, as disease yields and man comes to understand what causes him to age. This is the forecast for an age of peace" (Strauss 1954, 9).

emerges from experience and experiments. His primary goal was to correct misconceptions about science that derive from "the empiricist methodological tradition" of Bacon, Locke, Mill, and – most recently – Karl Pearson. The view that scientists collect, classify, and organize facts impersonally and without preconception or prejudice "is altogether wrong," Kuhn insisted, and his audience would come to understand that over the course of his lectures.

While introducing this methodological goal, Kuhn mentioned the institute's publicity:

> If any of you happens to have followed closely the advance notices of this series of lectures, you may have remarked that the topic just described bears very little relation to one announced in some of the flyers prepared by the Lowell Institute's copy writer. That topic was, I believe, described under a banner head reading: "What Are the Problems of Scientific Research Today?"
>
> I can scarcely imagine a more fascinating question; I should gladly attend a series of lectures devoted to it. *Except* that I doubt whether any serious student of science or scientific method would consider himself equipped to address such a subject. Therefore, with apologies for any confusion that the misrepresentation may have created, I should like to announce that I do not intend to deal with any of the problems raised by that question at any point in this series of lectures.[18]

Instead, he explained, "In the broadest and most fundamental sense, the objective of these lectures is to discover the relationship of this, the activity of the working scientist, to the products of his profession, to science as a body of human knowledge."

To make things clear, Kuhn specified further what he would not discuss. First, as a physicist, he would confine himself to issues in the history of physics, chemistry, and astronomy. It would be impossible to cover modern physical science, he noted, because it is "highly abstract and highly technical." But this was not a problem, he said, for the methodological conclusions to be drawn from historical material will nonetheless apply to and illuminate contemporary physical sciences.

Whether and how his conclusions would apply "to other sciences" outside of the physical sciences – social sciences – was another contentious topic. In their letter to President Truman, Conant and his colleagues warned that "it would be a serious mistake to include the social sciences"

---

[18] This and subsequent quotations from Kuhn's Lowell Lectures come from my forthcoming edition of the lectures. For unedited lecture scripts and outlines, see TSK Archives–MC240, box 3.

in the purview of the new foundation because administrators of basic science are not equipped to make parallel decisions about funding the social sciences.[19] In *On Understanding Science*, Conant noted that debate concerning similarities and differences between natural and social sciences was becoming more strident: "An increasing number of intelligent citizens believe that the social sciences rather than the physical or biological sciences hold the keys to the future. Others dissent violently from this view and even question the correctness of the phrase 'social science'."

All the more reason therefore that general education should examine the issues: "[I]f a layman is to have any clear ideas about the relation of the methods of physics, chemistry, and biology to education or to the investigations of current issues, he must understand the methods in question" (Conant 1947, 4, 5). Though Kuhn was speaking to laypeople in his lectures, he placed aside the issues orbiting the social sciences, noting that whether his conclusions apply "to other fields must finally be left to the man working in those fields."

Next, Kuhn's focus on science as a "body of human knowledge" meant he would not discuss technology and applied science. "We shall be concerned," he wrote

> with the sort of research the led to the Newtonian laws of motion, not with the manner in which these laws were applied in building new machines or instruments. We shall be concerned with the work of such men as Boyle and Dalton, insofar as this led to a new understanding and a new set of laws governing the formation of chemical compounds, but we shall not be concerned with the manner in which these laws, once arrived at and confirmed, were applied to the production of dyes, explosives, or plastics.

"Of course," he acknowledged, technology is an important component of the overall scientific enterprise. "But I have chosen to direct your attention and mine to the conceptual rather than the tangible aspects of scientific progress."

Finally, for a less visible but nonetheless revealing manifestation of Kuhn's aversion to these political topics, consider Kuhn's treatment of Pearson. Conant had introduced Pearson at the beginning of *On Understanding Science* and taken issue with his emphasis on impartial observation and "classification of facts" (Conant 1947, 14). Conant additionally rejected Pearson's contention that popular, state-supported education in science "is an education specially fitted to promote sound

---

[19] Reisch 2019b; New York Times, 1945.

citizenship" (Conant 1947, 6). In a detailed footnote, he argued that education in science's history, as opposed to science itself, would more effectively create "the basis for an improved type of citizenship" (Conant 1947, 111, n. 1; see also 10–11).

Kuhn also introduced Pearson in his introductory lecture, criticized his emphasis on the collection and classification of facts, and added that Pearson's image of "the scientific man" as objective and "unbiased by personal feelings" was deeply mistaken. Describing what Pearson had written, Kuhn's lecture script reads,

> "Such a man," he said, "... will scarcely be content with merely superficial statement, with vague appeal to the imagination, to the emotions, to individual prejudice; he will demand a high standard of reasoning, [and] a clear insight into facts and their results."

This is mistaken, Kuhn explained, because "prejudice and preconception are inextricably woven into the pattern of scientific research, and that any attempt to eliminate them would inevitably deprive research of its fruitfulness." Kuhn did not point out, however, that Pearson remarked about "the scientific man" to elaborate the social and civic importance he attached to the public understanding of science.[20] Kuhn indeed had a substantive methodological disagreement with Pearson about how research proceeds, and he dedicated his lectures to laying out his argument, but he silently separated that disagreement from a larger programmatic disagreement about whether and how the public should understand science.

3.6 Kuhn Theorizes a New Kind of Scientific Community

These restrictions of methodological scope remained in place when, about two years later, Kuhn began to draft outlines and notes for SSR. SSR took its familiar shape, however, only after another momentous development in Kuhn's theorizing about the nature of professional scientific communities – a turn that isolated further his methodological concerns from contentious social and political issues.

---

[20] Pearson had written (with those parts excised by Kuhn struck out), "such a man we may hope will carry his scientific method into the field of social problems. He will scarcely be content with merely superficial statement, with vague appeal to the imagination, to the emotions, to individual prejudice; he will demand a high standard of reasoning, a clear insight into facts and their results, and his demand cannot fail to be beneficial to the community at large" (Pearson, 11).

To appreciate Kuhn's innovation, consider the image of scientific communities implicit in Conant's philosophy of science and his general-education project. There, the boundaries separating educated citizens from professional scientists, while not transparent, were sufficiently translucent so that nonscientists and the public could come to understand "the tactics and strategies" of research through properly designed general education and, as I have argued, sufficiently translucent for citizens to recognize how the virtues of "civil courage" on display in the history of science circulate alike through a healthy democracy.

In SSR, scientific communities are comparatively closed to outsiders. This is a feature of "normal science," which makes possible the revolutionary model of science presented in SSR. For only by having strong, professional borders that screen out potentially subversive assumptions and premature theoretical exploration can a normal tradition sufficiently exhaust its internal resources and become a candidate for revolutionary change. As K. Brad Wray has rightly pointed out (Wray 2016), Kuhn's theory of normal science is not found in *On Understanding Science* and that I suggest is because of this underlying difference in conceptions of scientific community.

Evidence of this development in Kuhn's theorizing – and possibly the occasion for it – concerns correspondence to Kuhn from Philipp Frank, who himself championed Conant's general education program and campaigned regularly to increase the understanding of modern science on the part of the public and intellectuals in other fields. In late 1952, Frank invited Kuhn to join a research group he had formed to promote sociology of science. Kuhn responded by typing a long, critical letter that survives in multiple drafts, suggesting that Kuhn was searching for ways to specify what bothered him in Frank's presentation of the issues. Kuhn's main target was Frank's remark that one can readily distinguish occasions when scientists accept theories because they are "in agreement with observed facts" from occasions when sociological or "existential factors," as Frank put it, are also in play. Kuhn objected that Frank was oversimplifying matters and that it is not possible to distinguish these two cases because science is saturated with "existential" or "sociological" factors. Within any science, Kuhn explained, there is a professional "consensus" which influences "the sort of problems which a scientist considers worth attacking, the sort of experiment which he employs to resolve his problem, the abstract aspects of the experiment which he considers relevant to the solution, and the logical and experimental standards which he demands as 'proof.'"

The new research group should look at the sociological foundations of this consensus and, as he put it, "the ubiquitous role of the sociology of the professional group."[21]

These remarks may seem consistent with Conant's conception of scientific communities. If sociological forces circulate through science and society alike, those sociological dynamics might provide an entrée for laypeople to understand better how research proceeds, the kinds of pressures and anxieties professional scientists grapple with, and relations between scientific and civic values and virtues. But that is not what Kuhn meant. The "ubiquitous sociology of the professional group," as he put it, is ubiquitous only in professional science. It is different from the sociology of civic life and, as Kuhn noted, has little to do with "factors (like government, church, etc.) which at this time and place have relatively little impact upon decisions made by professional scientists about problems arising within their own sciences" and which exist "in analytic independence of the scientific enterprise" (ibid.).

A year and a half earlier, when Kuhn typed his joke about McCarthy, he evidently agreed with Frank that there were occasions when external social factors may substantively affect science's development. Now, it appears, Kuhn had decided against this possibility. He had spoken in his Lowell Lectures about the distinctive "phenomenal worlds," "perceptual worlds," and "behavioral worlds" in which scientists work, but in his response to Frank he offered a stronger, unified conception of communities that are closed to outside influence and operate according to their own sociological processes. Eventually, Kuhn would call these "normal" scientific communities.

3.7 SSR and Prospects for General Education

By the time he had written SSR, Kuhn had realized and articulated this theoretical development. Normal scientific communities are worlds unto themselves to the extent that "when paradigms change, the world itself changes with them," and intelligent and articulate scientists "will inevitably talk through each other when debating the relative merits of their respective paradigms" (SSR-4, 111, 109). If that is true, one might expect Kuhn to have grown skeptical about the promises Conant made on behalf of general education and its potential to enlighten the public. That is what we find in a talk he prepared in 1955, two years after Conant had left

---

[21] Kuhn, n.d., untitled document ("Dear Professor Frank"). TSK Archives – MC240, box 25, folder 53.

Harvard for Germany. The talk is titled "Can the Layman Know Science?"

He began by echoing Conant that the public *should* know more about science, for "in a democratic society" questions about funding and science policy inevitably involve "the citizen, the voter, the men and women who ultimately approve or disapprove policies." But the talk was largely pessimistic and dwelled on different factors behind the public's ongoing and "unprecedented ignorance" about science. Besides the rapidly increasing amount of knowledge (that made it impossible even for scientists to stay abreast of current developments) the public was hamstrung by misleading images of scientists in popular culture and advertising. The case studies Kuhn taught in *Natural Science 4*, he said, help dispel these stereotypes and cultivate a more nuanced understanding of creativity, complexity, and the "crooked paths" research often takes. But instead of declaring that this is at least a good start, Kuhn confessed at the end of his lecture that he did not know whether "the layman" can really come to know science – "at least not yet," his outline reads. And he implied that some other solution – besides Conant's – may be discovered "which can bridge at least part of the gap between science and the layman."[22] As "important" an issue as this was, Kuhn seemed drawn toward the view that Conant's program was not going to succeed.

3.8 Conclusion: A Politically Disengaged Historian and Philosopher of Science

Unlike Conant, Kuhn matured into a scholar for whom history and philosophy of science are disconnected from matters of civic and political concern. This does not mean, however, that Kuhn simply abandoned his childhood and teenage interests in politics and society. In his letter to Frank, for instance, as much as Kuhn isolated modern science from cultural, political, and sociology pressures of modern life, he theorized that similar dynamics and pressures operate *within* scientific communities, not as impediments but as functionally enabling factors. He drew upon considerations from pedagogy, psychology, epistemology, and linguistics to theorize how this "ubiquitous role of the sociology of the professional group" supports research and makes progress possible. As Kuhn put it in a letter to Charles Morris in 1953 when describing the monograph about scientific change he planned to write for the encyclopedia, "my basic

---

[22] Kuhn, 1955. TSK Archives – MC240.

problem is sociological" because any theory to be replaced during a scientific revolution "must be embedded in the professional group by which it will be overthrown."[23] The monograph that Kuhn completed several years later did not avoid political concepts so much as appropriate, transform, and rename them for Kuhn's now-unpolitical and scholarly purposes. The core concept of "paradigm," for example, evolved out of the overtly political "theories-as-ideologies" that Kuhn described in his letter to Morris and which appears in his notes and early drafts of SSR.[24]

In the 1970s, therefore, it may have seemed that Mr. Pietrzak had indeed asked Kuhn an odd question. For it was difficult to imagine that Kuhn's career and SSR itself first began to form in vital connection with geopolitically charged controversies about education, national character, and science funding in the early cold war. This suggests a more reflexive reading of Kuhn's response to Mr. Pietrzak. If we take seriously Kuhn's reputation as a revolutionary theorist who changed the scholarly understanding of science, and keep in mind SSR's remarkable account of historical myopia – the ways by which revolutionary texts render "invisible" the momentous crises, anxieties, and conflicts that precede their acceptance (SSR-4 chapter 11) – we can perhaps compare Kuhn to the revolutionary figures he described in his case studies. When he remarked to Mr. Pietrzak, "I cannot imagine what your informant had in mind," the McCarthy era had given birth to new scholarly norms and professional sensibilities very different from those that came before.

---

[23] Kuhn 1953.
[24] Another important example is Kuhn's provocative use of "dogma" – a use that extended no farther than his essay of 1963, "The Function of Dogma in Scientific Research" (Kuhn 1963). On this see Reisch 2019a, chapter 13. On Kuhn's use of "ideology," see Reisch 2019a, esp. chapters. 10–12. For a comparison of Kuhn's emphases on "conversion experiences" to contemporary interest in brainwashing, see Reisch 2012 and Reisch 2019a, chapter 6.

CHAPTER 4

Kuhn and Logical Positivism
On the Image of Science and the Image of Philosophy

J. C. Pinto de Oliveira

4.1 Introduction

Kuhn has been hailed as one of the main critics of logical positivism since the publication of SSR. But the "image of science" he criticizes in the book is not considered sufficiently clear. This has left room for doubt as to whether or not positivism, rarely directly cited, is a real target of Kuhn's criticism. However, in the first manuscript of the work, written in 1958, Kuhn contrasts science and art, and the traditional image of science is much clearer.

In an earlier paper on the first manuscript of Kuhn's *Structure*, I sought to show that in developing his philosophy of science Kuhn begins by sketching a traditional image of science by contrasting this image with the image of art. Then, by reducing the contrast, he brings science closer to art and thus proposes a new image of science (see Pinto de Oliveira 2017). But, as I have argued earlier, Kuhn also refers in this context to disciplines other than art. In the final version of SSR, he refers to politics (SSR-4, 92–94) and, in other texts, to philosophy and social sciences, as well as to literature (see, e.g., RSS, 136–138 and 216–217).[1] According to Kuhn, all these disciplines could function as a model for better understanding the development of science.

In this work, I intend to investigate the relationship between science and philosophy. The project is not to compare science with all these other disciplines, starting first with philosophy. The study of the relation between science and philosophy reinforces what I have said earlier about science and art. In addition, it allows me to try to elucidate the issue of the relation between Kuhn and logical positivism.

---

[1] In the first manuscript, Kuhn says that art is the prototype of noncumulative disciplines (see Pinto de Oliveira 2017, 761, note 28).

As my earlier paper on Kuhn's manuscript revealed, the traditional image of science clearly appears in the metaphor used by Carnap in 1928 on the bricks of scientific construction (Pinto de Oliveira 2017, 755). What I want to show here is that this traditional image is present throughout the work of the logical positivists and that, therefore, positivism has its share of responsibility in the image of science that Kuhn criticizes. And one might say that it is a significant share, as in the case of Sarton and Conant (see Pinto de Oliveira 2017, 751–754), because logical positivism is a contemporary movement and still preserves that image (see SSR-4, 98).

The strategy of the chapter is very simple. I will argue that the traditional image of science that logical positivism presents in contrast to philosophy – particularly in the context of its proposed scientific philosophy (Section 4.2)[2] – is the same one that Kuhn presents through his contrast with art and, more sparsely, with philosophy itself (Section 4.3). Thus, in the final section, it will be possible to adequately measure the share of responsibility that belongs to logical positivism with respect to the traditional image of science and this will give us a better understanding of Kuhn's relation to positivism.[3]

4.2 The Project of a Scientific Philosophy and the Logical Positivist Image of Science

The difficulty in showing the relation of logical positivism to the traditional image of science, an image that focuses on the idea of cumulative progress, is that the progress of science, or the so-called dynamics of science (see ET, 267), seems neither to be a Carnapian nor a positivist idea. As Hempel writes, "[T]he analytic empiricist school was not much concerned with the analysis of theoretical *change*" (Hempel 2001, 365). And this seems very clear in Carnap's last book, directly devoted to the philosophy of science: *An Introduction to the Philosophy of Science*. There are very few references to topics such as scientific revolution, change of theories or progress, and none of these terms appear in the index (see Carnap 1966/ 1995). And the same goes for Nagel (1961).

So if there are few references to cumulative progress in the work of logical positivism it is because there are few references to the very notion of progress.

---

[2] I refer here generically to logical positivism because I believe that the proposal of a scientific philosophy and the traditional image of science are shared by several members of the movement, as I point out later. It does not mean, therefore, the denial of individual differences on this and other subjects. I also make no distinction here between logical positivism and logical empiricism, in the same way as occurs, for example, in Richardson and Uebel (2007, 1, note 1).

[3] On the relations between Kuhn and logical positivism, see also Pinto de Oliveira 2007 and 2015.

The nature of relations between successive scientific theories is explained by the logical positivists and sympathizers of the movement through what they call the reduction of theories (see, e.g., Nagel 1961, chapter 11).

As Frederick Suppe suggests, the reduction theory can be understood as logical positivism's theory of scientific development (Suppe 1974, 53–56). In other words, the theory of reduction, although not part of the standard set of theses which he calls "the Received View," would be an adequate way of explaining the admitted progress of science. It might be said, using Carnap's terms, that the theory of reduction would provide an *explicatum* for the vague, implicit idea of scientific progress according to the logical positivists, which would be the *explicandum* (see Carnap 1950, 3).

In fact, I am not interested here in the theory of reduction. The *explicandum* as such is sufficient for my purposes. What I intend to argue here is that there is a simpler way of showing the commitment of logical positivism to cumulative progress and the traditional image of science. This can be done through the positivist conception or proposal of a scientific philosophy. Embedded in it, there is an image of science, which I will try to reveal. As we shall see, it stands out particularly for its characteristic progress. It is essentially due to its success, to progress, that science becomes a model for philosophy. With their proposal of a scientific philosophy, the logical positivists intend that this progress be extended to philosophy.[4]

In a retrospective analysis of the philosophical project of logical positivism, a project with which he himself was involved, Charles Morris wrote in 1963: "[S]een in perspective, science has been the norm or model controlling the entire development of logical empiricism ... In this way, as Carnap made clear in the Foreword to *Der logische Aufbau der Welt*, it was hoped to build by cooperative work a philosophy that could advance as science advances." (Morris 1963, 96)

What Carnap writes in this passage referred to by Morris – giving account of what would occur in science and in a scientific philosophy – deserves to be quoted in full here:

> The new type of philosophy [scientific philosophy] has arisen in close contact with the work of the special sciences, especially mathematics and physics. Consequently they have taken the strict and responsible orientation of the scientific investigator as their guideline for philosophical work, while the attitude of the traditional philosopher is more like that of a poet. This

---

[4] The proposal of a scientific philosophy is an essential part of the logical positivist project. Alan Richardson considers it "the project's primary philosophical concern" (Richardson 2008, 90). See also Carnap in Reichenbach 1959, vii).

new attitude not only changes the style of thinking but also the type of problem that is posed. The individual no longer undertakes to erect in one bold stroke an entire system of philosophy. Rather, each works at his special place within the one unified science. For the physicist and the historian this orientation is commonplace, but in philosophy we witness the spectacle (which must be depressing to a person of scientific orientation) that one after another and side by side a multiplicity of *incompatible philosophical systems* is erected. If we allot to the individual in philosophical work as in the special sciences only a partial task, then we can look with more confidence into the future: in slow careful construction insight after insight will be won. Each collaborator contributes only what he can endorse and justify before the whole body of his co-workers. *Thus stone will be carefully added to stone and a safe building will be erected at which each following generation can continue to work.* (Carnap 1928/1967, xvi–xvii, emphasis added)[5]

This perspective on science and philosophy is thoroughly shared by other logical positivists. Reichenbach, for example, in a 1931 text, emphasizes the positive process of specialization in philosophy, since it would follow the track of scientific specialization. He writes: "The process of differentiation constitutes a transition from the intuitive to the scientific method, from isolated speculation to scientific co-operation. Philosophy changes from being a sweeping world view to becoming a progressive science." (Reichenbach 1931, 83)

Neurath, in 1937, stresses the capacity for communication among scientists, "no matter how different their hypotheses and theories may be," and associates it with the project of a scientific philosophy. He says:

A new kind of cooperation is becoming possible, related to the kind we have known for a long time in the special sciences. Biologists can communicate with one another about the results of their research as profitably as geologists, or as astronomers who study cosmic rays, no matter how different their hypotheses and theories may be. They differ in this from philosophers, whose different schools are closed to one another. A phenomenologist will find it hard to converse with a speculative idealist, and both of them will find it hard to converse with a critical idealist. (Neurath 1937/1987, 137)

Neurath also draws attention in a previous text to the "republic of letters" in which scientists coexist democratically and to which he opposes the feudal lords of philosophy:

Strict self-control leads to successful cooperation between scientific specialists in the most diverse fields. *Metaphysical terms divide – scientific terms unite.*

---

[5] I believe "knowledge" is a better translation than "insight" for the word "Erkenntnis," originally used by Carnap in German. It is how the term was translated, for instance, in Galison (1990, 733).

> Scientists, united by a unified language, form a kind of workers' republic of letters, no matter how much else may divide them as men. Philosophers on the other hand are comparable to the feudal lords of San Gimignano. They sit in their lonely towers in the dark of night and seek to guard themselves against their neighbours by raising their towers ever higher and higher. (Neurath 1932/1987, 23)

And he writes, now along with Hahn and Carnap, in the so-called manifesto of logical positivism or of the Vienna Circle (1929): "This circle has no rigid organization; it consists of people of an equal and basic scientific attitude; each individual endeavours to fit in, each puts common ties in the foreground, none wishes to disturb the links through idiosyncrasies. In many cases one can deputise for another, the work of one can be carried on by another." (Neurath et al. 1929/1973, 299)

Of course, Schlick himself, to whom the manifesto is dedicated, thinks the same. He writes in 1931:

> [E]very thinker [in traditional philosophy] seeks his own foundation and does not wish to stand on the shoulders of his predecessors. ... [P]ractically all great thinkers have sought for a radical reform of philosophy and considered it essential. This peculiar fate of philosophy has been so often described and bemoaned that it is indeed pointless to discuss it at all. Silent scepticism and resignation seem to be the only appropriate attitudes. Two thousand years of experience seem to teach that efforts to put an end to the chaos of systems and to change the fate of philosophy can no longer be taken seriously.

He adds, not surprisingly:

> I am convinced that we now find ourselves at an altogether decisive turning point in philosophy, and that *we are objectively justified in considering that an end has come to the fruitless conflict of systems*. We are already at the present time, in my opinion, in possession of methods which make every such conflict in principle unnecessary. What is now required is their resolute application. (Schlick 1931/1959, 53–54, emphasis added)

And it is worth quoting Carnap's curious text of 1931 in which the epistemological privilege of science in the face of philosophy is highlighted very clearly:

> It would be conceivable that each person could make his protocol sentences agree with those of others only with great difficulty or not at all ... luckily, in fact we find ourselves in a position to bind together our protocols with those of a hundred other people in a common elaboration. If someone appears who, on the basis of his protocols, builds a science that is not consistent with the one constructed by our hundred people, then we vote

> him down; we say of him (depending on the circumstances) that he is colorblind, or a poor observer, or a dreamer, or a liar, or a madman. If one found that against our one hundred there is another one hundred with a common science that cannot be unified with ours, then we couldn't vote them down. In case further research would not lead to agreement we should accept the fact that *different groups possess unalterably diverse scientific systems. Luckily, this is not the case.* (*Apud* Coffa 1977, 217, emphasis added)

By the way, Karl Popper, even though he should not be considered a logical positivist, paints more dramatically the picture dominated by Carnap's supposedly "unalterably diverse scientific systems." According to him, "it would amount to a new 'Babel of Tongues'" and "the soaring edifice of science would soon lie in ruins" (Popper 1959/1968, 104 – section 29).

Despite the great energy expended in the project of a scientific philosophy, it did not, as we know, lead to the expected results (see Rorty 1982, 215–216). Even so, the project was not expunged from the positivist program in its so-called mature phase. This directly follows from the publication of Reichenbach's book *The Rise of Scientific Philosophy* (1951). But I will not focus here on this text, although it is the most specific, in order to add further evidence. Reichenbach writes, for example, in a 1948 article in which he relates scientific philosophy and the history of philosophy:

> Those among us who have taught one of the sciences will know what it means to teach on a common ground. The sciences have developed a general body of knowledge, carried by universal recognition, and he who teaches a science does so with the proud feeling of introducing his students into *a realm of well-established truth*. Why must the philosopher renounce the teaching of established truth? ... Imagine a scientist who were to teach electronics in the form of a report on views of different physicists, never telling his students what *are* the laws governing electrons. The idea appears ridiculous. Though the physicist does mention the history of his field of study, the views of individual physicists appear as *contributions to a common result established with a superpersonal validity and universally accepted.* Why must the philosopher forgo a generally accepted philosophy? (Reichenbach 1948, 135–136, emphasis added)

In his "Introductory remarks" to the English edition of Reichenbach's *The Philosophy of Space and Time* (1957), Carnap writes:

> Even more outstanding than the contributions of detail in this book is the spirit in which it was written. The constant careful attention to scientifically established facts and to the content of the scientific hypotheses to be analyzed and logically reconstructed, the exact formulation of the philosophical results, and the clear and cogent presentation of the arguments

supporting them, make this work a model of scientific thinking in philosophy. (Carnap 1957, vii; see also Carnap in Reichenbach 1959, vii)

"The spirit" in which the book was written, to which Carnap refers, is highlighted in Reichenbach's original Introduction (1928). According to him, scientific philosophy is "a product of teamwork":

> This accumulation of common knowledge is the characteristic mark of the new philosophical orientation, which, due to its origin in the empirical sciences, stands even methodologically in contrast to the isolated systems of the speculative philosophers and gains its superiority from this source. Philosophy of science is not intended to be one of those systems that originate in the mind of a lonely thinker and stand like marble monuments before the gaze of generations, but should be considered a science like the other sciences, a fund of cooperatively discovered propositions whose acceptance, independent of the framework of a system, can be required from anybody interested in these matters. The meaning of concepts may vary, of course, depending upon the context in which they are used; but this kind of ambiguity can be avoided by making language more precise and need not lead to a renunciation of objective philosophical knowledge altogether. If the effect of the philosophy of systems was to destroy the concept of philosophical truth and replace it by the concept of consistency within the system, one may see as the noblest aim of scientific philosophy the establishment of the concept of objective truth as the ultimate criterion of all philosophical knowledge. (Reichenbach 1928/1957, xv–xvi)

And in his intellectual autobiography, published after Kuhn's SSR, Carnap compares philosophy and science through their historiographies (see Pinto de Oliveira 2015, 219). He writes:

> The task of the history of philosophy is not essentially different from that of the history of science. The historian of science gives not only a description of the scientific theories, but also a *critical judgment of them from the point of view of our present scientific knowledge*. I think the same should be required in the history of philosophy. This view is based on the conviction that *in philosophy, no less than in science, there is the possibility of cumulative insight and therefore of progress in knowledge*. This view, of course, would be rejected by historicism in its pure form. (Carnap 1963, 41, emphasis added)

To summarize, we can say that the essential aspects that the logical positivists emphasize in the traditional image of science in contrast to traditional philosophy are the following[6]:

---

[6] All references are from the works and passages quoted before in this section.

A. Science progresses (it is cumulative and progressive) and philosophy does not (Carnap 1928/1967; Reichenbach 1957; Carnap 1963).
B. Science is a collective and cooperative enterprise and philosophy is an individual and solitary activity (Reichenbach 1931; Neurath 1932 and 1937; Neurath 1973; Schlick 1931/1959; Reichenbach 1928/1957).
C. Scientific theories are compatible with each other while philosophical systems are incompatible (Coffa 1977; Carnap 1967; Neurath 1932/1987 and 1937/1987; Schlick 1959; Carnap 1963).
D. Science is empirical and has an intersubjective character, while philosophy is speculative and has a subjective character (Reichenbach 1931; Neurath 1932/1987 and 1937/1987; Reichenbach 1948; Reichenbach 1928/1957; Coffa 1977).

4.3 Kuhn and the Traditional Image of Science: The Contrast with Art

For an immediate comparison to what has just been pointed out earlier, we might say that the essential aspects of the traditional image of science that Kuhn highlights in contrast with art are A and C: **A.** Science progresses (it is all the way cumulative and progressive) and art does not progress. **C.** Scientific theories are commensurable while the artistic styles are incommensurable.[7]

I emphasize these aspects of the relation between science and art based especially on Pinto de Oliveira (2017). There I examined *Structure*'s first manuscript, written in 1958, to show that Kuhn, at that moment, considered the comparison between the traditional images of science and art as the most appropriate way to announce his project: to change the image of science by bringing it closer to the image of art. As I try to argue there, this recourse to the history of art is not merely an occasional one but one that allows us to understand Kuhn's retrospective statement, as concise as it is striking, that SSR was a "belated product" of his discovery of the parallels between science and art (ET, 340).

As my article developed the theme of the relation between science and art, with its varied nuances, and even transcribed passages of the unpublished text of Kuhn, I will resume the discussion here in a more

---

[7] The logical positivists do not use the word "incommensurable." In the final section, I will argue that they intuitively conceive the idea of incommensurability when talking about the relationship between philosophical systems, just as Kuhn does with respect to art and philosophy itself. Kuhn's originality lies in seeking to develop the concept and apply it to science.

direct way.[8] Resorting to the article and other texts, I limit myself to sketching the two aspects I pointed out earlier. I take these to be the main aspects of the traditional image of science that Kuhn highlights through the contrast with art.

4.3.1 Science Progresses and Art Does Not Progress

From the outset of the manuscript, Kuhn compares the development of science and art to mark the difference between them, according to the traditional conception. He sketches what the cumulative character proper to science would be, the character which opposes it to art:

> Both disciplines display continuity of historical development ... yet the relation of present to past in these two fields is clearly distinct. Einstein or Heisenberg could, we feel sure, have persuaded Newton that twentieth-century science has surpassed the science of the seventeenth century, but we anticipate no remotely similar conclusion from a debate between, say, Rembrandt and Picasso. In the arts successive developmental stages are autonomous and self-complete: no obvious external standard is available for comparisons between them. (Kuhn (n.d.), M1, TSK Archives – MC240, box 4, folder 3, (2–3), quoted in Pinto de Oliveira 2017, 749)

The traditional image of the development of science also manifests itself through many converging metaphors. In Pinto de Oliveira (2017, 754) I present several of them, due mainly to Sarton or revisited by him, such as that of the mountain, the dwarfs on the shoulders of giants, the giant that grows indefinitely, the immortal tree, a ladder or a ramp. But the metaphor that Kuhn highlights in *Structure*'s manuscript to account for the traditional image of science – as opposed to the image of art – is that of the scientific building, which is built cumulatively over time: "an ever-growing edifice to which each scientist strives to add a few stones or a bit of mortar" (Kuhn (n.d.), M2, TSK Archives – MC240, box 4, folder 3, (2), quoted in Pinto de Oliveira 2017, 748).

The metaphor of the building, as I point out in the 2017 article (755), appears in several authors such as Sarton, Zilsel, Popper and Carnap himself. Kuhn uses it to mark his position against the image of science

---

[8] I use Kuhn's manuscript here, sometimes quoting its passages only partially, according to the conveniences of argumentation and availability of space. In Pinto de Oliveira 2017, the previously unpublished passages are comprehensively transcribed. As in the article, I refer here to the first manuscript as M1 (Kuhn (n.d.), M1, TSK Archives – MC240, box 4, folder 3) and to the second as M2 (Kuhn (n.d.), M2, TSK Archives – MC240, box 4, folder 3).

being associated to it. In a passage from the second manuscript he makes it clear that we must think of the process of building science in other terms. His observations, which already point to the direction of a new image of science, also contribute to a better understanding of the traditional image:

> [W]e may have to recognize that the addition of new bricks demands at least partial demolition of the existing structure, and that the new edifice erected to include the new brick is not just the old one plus, but a new building. We may, that is, be forced to recognize that new discoveries and new theories do not simply add to the stock of pre-existing scientific knowledge. They change it. (Kuhn (n.d.), M2, TSK Archives – MC240, box 4, folder 3, (7), quoted in Pinto de Oliveira 2017, 756)[9]

Besides bringing science closer to art, Kuhn also brings art closer to science, reducing the contrast between them in both directions. According to him, the cumulativity observed in science also exists in the arts.[10] Thus, as he says, "if cumulativeness is to distinguish the developmental pattern of science and art, that cannot be because art is never cumulative but because science is always so" (Kuhn (n.d.), M1, TSK Archives – MC240, box 4, folder 3, (4), quoted in Pinto de Oliveira 2017, 750).

In a 1969 text, reissued in 1977, Kuhn raises this idea by challenging what he considers an "often repeated generalization": "To say with pride, as both artists and scientists do, that science is cumulative, art not, is to mistake the developmental pattern in both fields" (ET, 350).

4.3.2 Scientific Theories Are Commensurable and Artistic Styles Are Incommensurable

Kuhn introduces the concept of incommensurability in *Structure*'s first manuscript. Significantly, he uses it first in the case of art, stating that the transition between one stage of artistic development and another is a transition between incommensurables (see Pinto de Oliveira 2017, 749). On the other hand, Kuhn says, "[T]he present state of science always seems to embrace its past stages as parts, which is what the concept of cumulativeness means". And he adds that "we see in the development of science no equivalents for the total shift of artistic vision – the shift from one integrated set of problems, images, techniques, and tastes to another"

---

[9] In a text from his graduate student time, Kuhn himself uses the metaphor in the traditional manner (see Marcum 2015b, 30).
[10] As he says in SSR-4: "For many centuries, both in antiquity and again in early modern Europe, painting was regarded as the cumulative discipline" (160). See also Section 4.4.

(Kuhn (n.d.), M1, TSK Archives – MC240, box 4, folder 3, (3). See the entire passage in Pinto de Oliveira 2017, 749).

For an adequate understanding of what Kuhn has in mind when he speaks of this "total shift of artistic vision," which will serve as a model for the conception of scientific revolution, we can draw upon the historian of art Ernst Gombrich, referred to by Kuhn himself (ET, 340). Here I will briefly follow the line of development Gombrich presented in *The Story of Art*.[11] This will allow us to identify the presence of some fundamental notions of Kuhn's philosophy of science, such as the concepts of paradigm and incommensurability, as he thinks of them within the history of art.

When looking at, for example, the reliefs and paintings of Egyptian artists, says Gombrich, we are most likely to find them strange. And, as he points out, "perhaps this is connected with the different purpose their paintings had to serve" (Gombrich 1950/1995, 60). Obeying strict rules, they represented, characteristically, the human head always in profile, but at the same time with the eyes as seen from the front. This Egyptian style was more conceptual and thus profoundly differed from Greek art. It was in Athens, in the sixth century BC, that, according to Gombrich, "the greatest and most astonishing revolution in the whole history of art" occurred with the discovery of foreshortening and perspective (Gombrich 1950/1995, 77). The traditional formulas for representing the human body were no longer in vogue. Every Greek artist "wanted to know how *he* was to represent a particular body. The Egyptians had based their art on knowledge. The Greeks began to use their eyes" (Gombrich 1950/1995, 78).

A revolution of the same content happened again in Italy in the fourteenth century. According to Gombrich, primitive Christian art, before Giotto, had reverted to certain standards of Egyptian art. And the new perspective culminated with the Renaissance in the sixteenth century, when it was admitted that painting had reached perfection with artists like Michelangelo, Raphael, and Leonardo.

But would it make sense to ask whether Michelangelo was a better painter than a thirteenth-century Christian artist, or whether the Greeks were better painters than the Egyptians, or even, to use Kuhn's example in

---

[11] I follow Gombrich's footsteps in *The Story of Art* more thoroughly in Pinto de Oliveira 2014. See also Pinto de Oliveira 2017, 760, where I call attention to the influence of this book on Kuhn's conception about the relation between science and art. By the way, in Agassi and Jarvie 2008 (164, note 17), the authors write that "Gombrich began work on *The Story of Art* (1950) before he was familiar with Popper's views on rationality. *Art and Illusion* (1960) made use of what he learned from Popper."

Structure's first manuscript, whether Picasso, specifically the cubist Picasso, was better than Rembrandt? In art, it seems absurd to compare two painters of such different styles as Rembrandt and Picasso and ask which of them would be the best. They had very different and incommensurable conceptions of art. There is no single criterion, no common denominator, that would allow us to directly evaluate the quality of one in relation to the other.

The case of science, as it is traditionally supposed, would be completely different. As Kuhn writes, in science "problems seem to be set by nature and in advance, without reference to the idiom or taste of the scientific community" (Kuhn (n.d.), M1, TSK Archives – MC240, box 4, folder 3, (3)). And he adds, "Apparently, therefore, successive stages of scientific development can be evaluated as successively better approximations to a full solution" (Kuhn (n.d.), M1, TSK Archives – MC240, box 4, folder 3, (3). See the entire passage in Pinto de Oliveira 2017, 749). That is, we can say that within science the comparison would be much simpler, as if it were always performed within the same style.

For Kuhn, however, the history of science denies support for this kind of relation between theories imagined by earlier philosophers, such as the logical positivists. According to him, the rival theories differ from one another in a more radical way than a long tradition made us believe. On some occasions, the decision to be made by scientists involves a choice between alternative forms of conceiving an object and doing science as if the competing theories corresponded, as a matter of fact, to different and incommensurable scientific "styles" or, as Kuhn says, different paradigms.

Michelangelo's contemporaries thought he was a better sculptor than the Greeks. This comparison was possible because the works of Michelangelo and those of the Greeks could be considered manifestations of the same style. After all, the so-called Renaissance in the arts was, precisely, a resumption of Greek and Roman art. On the other hand, we saw that it would be much more complex to compare the Greek artist with the Egyptian, or Rembrandt with the cubist Picasso, to know which would be the best. And so, in the same way, the choice between competing scientific theories or paradigms could be as complex as between different styles in art, as in the case Kuhn says that there is ipso facto a scientific revolution.

To conclude the comparison, there remains an observation on the aspects B and D of the logical positivists' image of science. Kuhn does not emphasize these aspects of the traditional image of science in contrast to art in *Structure*'s first manuscript. But, certainly, they are subsumed by what he calls "the classic dichotomies" present in the contrast between science and art. In the same place, cited earlier, in which he says that the book is a belated product of the parallel between science and art, he writes that SSR is concerned "to deny, at least by strong implication, that art can readily be distinguished from science by application of the classic dichotomies between, for example, the world of value and the world of fact, the subjective and the objective, or the intuitive and the inductive." (ET, 340)

So it can be said that the traditional image of science outlined by the logical positivists in contrast with philosophy and by Kuhn in contrast with art is essentially the same. In the next section, I will better adjust the focus to make this convergence even clearer.

4.4 Final Comments

The aspects of the image of science present in the proposal of a scientific philosophy allow us to understand why the progress of science is not properly a theme for logical positivism. The scientific cumulative progress, stated in item A, is taken for granted and is supposed to come naturally from the other three items. In summary, science is empirical and intersubjective; scientific theories are compatible with each other; scientific activity is collective and cooperative; therefore, science presents cumulative progress, unlike philosophy.

As we have seen, the outline of the new image of science in the first manuscript is through the reduction of the high contrast between science and art. For Kuhn art is cumulative in the same way that science is noncumulative (with due caveats). By reducing the contrast between the poles, the very notion of progress is problematized, and the progress of science, which is now close to the progress of art and hence away from the traditional conception, becomes a key issue to be faced in Kuhn's philosophy of science.

In fact, what has been said about science and art is also true for philosophy, which particularly interests us here. Kuhn does not refer extensively to the relation between science and philosophy but does so in the same sense of the relationship between science and art. As he says in the published version of SSR: "The theologian who articulates dogma or the philosopher who refines the Kantian imperatives contributes to progress ... The man who argues that philosophy, for example, has made no progress emphasizes that there are still Aristotelians, not that Aristotelianism has failed to progress." (SSR-4, 162)

In a 1965 text, reissued in ET, Kuhn addresses Popper to point out the proximity between science and philosophy: "Only when they must choose between competing theories do scientists behave like philosophers. That, I think, is why Sir Karl's brilliant description of the reasons for the choice between metaphysical systems so closely resembles my description of the reasons for choosing between scientific theories." (ET, 273) And in a 1969 text reprinted in RSS, Kuhn speaks at the same time and in the same sense of the relation among science, art, and philosophy:

> Consider for a moment the evolution of philosophy or of the arts since the end of the Renaissance. These are fields often contrasted with the established sciences as ones which do not progress. That contrast cannot be due to the absence of revolutions or of an intervening mode of normal practice. On the contrary, long before the similar structure of scientific development was noticed, historians portrayed these fields as developing through a succession of traditions punctuated by revolutionary alterations of artistic style and taste or of philosophical viewpoint and goal. (RSS, 137. See the entire passage, 137–141)

Thus, as one can see, Kuhn reduces the contrast between science, on the one hand, and art, philosophy, and other disciplines on the other, conceiving both for science and for other disciplines normal and extraordinary periods throughout their historical development. The traditional image, one might say, establishes its high contrast by taking science and art (or philosophy) as two poles. At one end we have science, with only one (normal) period, and at the other, art and philosophy, with a succession of extraordinary periods. This is the "translation" into the Kuhnian language of the traditional idea that, among other things, as we have seen, science progresses (it is strictly cumulative) and art and philosophy do not.

To conclude, I emphasize the convergence of the contrast between science and philosophy as made by the logical positivists and science and art (as well as philosophy and other disciplines) as made by Kuhn. And it can be clearly seen that the logical positivist image of science is the same traditional image of science that Kuhn contrasts with art, already present in Conant and Sarton (see Pinto de Oliveira 2017, 751–755). As we have seen, evidence for this can be found in the metaphors employed and even in the form chosen in both cases to establish the contrast between the disciplines.

In my earlier article on *Structure*'s first manuscript, I quote a long passage from Conant, which I reproduce partially here, since he refers to science, art, and philosophy:

> Bring back to life the great figures of the past who were identified with the subjects in question. Ask them to view the present scene and answer whether or not in their opinion there has been an advance. No one can doubt how Galileo, Newton, Harvey, or the pioneers in anthropology and archeology would respond. It is far otherwise with Michelangelo, Rembrandt, Dante, Milton, or Keats. It would be otherwise with Thomas Aquinas, Spinoza, Locke or Kant. We might argue all day whether or not the particular artist or poet or philosopher would feel the present state of art or poetry or philosophy to be an advance or a retrogression from the days when he himself was a creative spirit. There would be no unanimity among us. (Conant 1957, 34)

As for the relation between science and philosophy, this passage shows that Conant and the logical positivists share the traditional image of science, which they oppose to the image of philosophy. The difference refers to those authors that the logical positivists consider scientific philosophers such as Kant. Reichenbach, for instance, begins his discussion of Kant's epistemological ideas by recognizing Kant's importance, which would come precisely from the fact that he could be considered a scientific philosopher (*avant la lettre*). Reichenbach says that "what makes Kant's position so strong is its scientific background" (Reichenbach 1951, 42). And he writes later:

> The history of philosophy, which up to the time of Kant manifested itself in the form of philosophical systems, should be regarded as continued after Kant not by the pseudosystems of the imitators of a great past, but by the new philosophy that grew out of the science of the nineteenth century and was continued in the twentieth century. (Reichenbach 1951, 122)

This new philosophy of which Reichenbach speaks culminates in the very logical positivism, which has Kant as the historical interlocutor and not the so-called German idealism. As Reichenbach says, German idealism constitutes a rupture in the argumentative line that takes science as the model of knowledge for epistemological reflection. And Carnap argues that German idealism did not give due importance to science (Reichenbach 1951, 122–123; Carnap 1966/1995, 12).

As a general conclusion to his evaluation of Kant as a philosopher, Reichenbach writes:

> *Had Kant lived to see the physics and mathematics of our day he might very well have abandoned the philosophy of the synthetic a priori.* So let us regard his books as documents of their time, as the attempt to appease his hunger for certainty by his belief in the physics of Newton ... This origin explains the system's success and its failure, explains why Kant has been regarded by so

many as the greatest philosopher of all time, and why his philosophy has nothing to say to us who are witnesses of the physics of Einstein and Bohr. (Reichenbach 1951, 44, emphasis added).

Reichenbach uses the same kind of "thought experiment" as Conant and Sarton, which Kuhn also uses in *Structure*'s first manuscript to outline the traditional image of science. In Reichenbach's hands, the "experiment" brings Kant into the present and transforms him into a logical positivist.[12] He does this with the *scientific* philosopher Kant in the same way that Carnap does with Aristotle, bringing him into the present and making him a contemporary scientist: "Indeed, he would probably be one of today's leading scientists" (Carnap 1966/1995, 119).

What is important in Kuhn's work, as he himself suggests (SSR-4, p. 207 and ET, 150–151), is his controversial extension to science, especially to physics and chemistry, of the conception of development observed in art, philosophy, and other fields. In a feedback process Kuhn starts from more intuitive notions of "paradigm" and "incommensurability," as found in history of art, as well as history of philosophy. He develops them within his philosophy of science and thus, at a higher level of conceptualization, they later attract the interest of virtually all areas of culture (see Pinto de Oliveira 2017, 761).

And I believe that logical positivism also has an intuitive notion of incommensurability, when considering the relationship among the various systems of traditional philosophy. We have seen in Section 4.2 how Neurath draws attention to the "republic of letters" in which scientists coexist democratically and to which he opposes the "feudal lords" of philosophy, locked and solitary in their towers (Neurath 1932/1987, 23). And so does Carnap when he refers to incompatible philosophical systems to which he opposes the unique edifice of science, constructed by carefully placing one stone upon another, and "at which each following generation can continue to work" (Carnap 1928/1967, xvi–xvii).[13]

One might even think that Carnap would have announced the concept of incommensurability in science in 1931 (as Alberto Coffa suggests) by

---

[12] Carnap writes: "[A]s Moritz Schlick once remarked, empiricism can be defined as the point of view that maintains that there is no synthetic a priori. If the whole of empiricism is to be compressed into a nutshell, this is one way of doing it" (Carnap 1966/1995, 180).

[13] The way in which Carnap and Reichenbach refer to German idealism, as we saw earlier, contrasting it with the Kantian system – which would give due importance to science – suggests the presence of incommensurability between them. We can say that the two systems, like Kuhn's paradigms, "lack a common measure because they use different concepts and methods to address different problems, limiting communication across the revolutionary divide" (Oberheim and Hoyningen-Huene 2018, 2).

speaking about different groups that would have "unalterably diverse scientific systems" (*Apud* Coffa 1977, 217, quoted in section 2). Carnap uses an intuitive notion traditionally associated with philosophy, the idea of incompatibility between philosophical systems (and he uses the term "systems" in the case of science as well), just as Kuhn initially identifies, more intuitively, the notion of incommensurability in the relationship between styles in art.

In a "thought experiment" merely rhetorical, Carnap imagines the possibility of incommensurability in science precisely to deny it and to praise the happy circumstance in which we are regarding scientific knowledge. That is, according to Carnap, fortunately, there is no incommensurability in science. Thus, for logical positivism and its project of a scientific philosophy, originating in the 1920s and 1930s and, as we have seen, still present in the 1950s and 1960s, scientific theories are cumulative and compatible, unlike philosophical systems. And we can also say that they are commensurable, unlike philosophical systems, as suggested by Carnap himself when he rejects the idea of "unalterably diverse scientific systems."

To conclude, for Kuhn, the process by which philosophy and other disciplines develop is a model for understanding the process by which science develops. Kuhn offers an alternative to the traditional image of the cumulative progress of science, present in science textbooks, and also, as we have seen here, in many texts of the logical positivists: the process understood as "a succession of tradition-bound periods punctuated by non-cumulative breaks" (SSR-4, 207). In this strict sense, it can be said that Kuhn confronts the logical positivists' *scientific philosophy* with the idea of a *philosophical science* – something very distinct and certainly incommensurable with the positivist conception.[14]

The idea of philosophical science also means that philosophy or metaphysics has not lagged behind, as an outdated stage of the development of knowledge (as Comte and the positivists put it), but that science is embedded in philosophy – not in its more pragmatic institutional aspects, but from an epistemological point of view. This is also what Koyré thinks, as can be seen in Gerard Jorland's book, significantly entitled *La science dans la philosophie* (see Pinto de Oliveira 2020, 392).

---

[14] Kuhn suggests that his philosophy of science and that of Popper are incommensurable (RSS, 124–125) and the same would certainly be true of its relation to logical positivism. Thus, one should not expect of Kuhn a "point by point" criticism of logical positivism, as Popper expected of Adorno (see Popper et al. 1969/1976). In SSR, Kuhn refers directly to logical positivism only on pages 98 and 101. But what is at stake in the relationship between incommensurables is what Kuhn called "total change of vision," whether it occurs in science, art, or philosophy (see earlier, and also SSR-4, 149).

I believe that the very "thought experiment" of reviving a scientist (or a scientific philosopher) today and claiming that he would agree with the contemporary conception seems to reveal some assumptions of logical positivism with which Kuhn would not agree. The problem lies in the idea of reviving one single individual, separated from his circumstances, from his scientific community, and inserting him loose into the contemporary context. The experiment is carried out as if an individual decision were self-sufficient. This possibility would deny the Kuhnian thesis according to which scientific knowledge is intrinsically or necessarily social and not only customarily or contingently social (see, e.g., ET, xx).[15] But this would be a subject for an upcoming work.

Acknowledgment

I am grateful to all those who commented on earlier drafts of this work, especially Amelia Oliveira, Luisa Dezopi, and the editor K. Brad Wray. I would also like to thank Baruana Calado for translations and revisions.

---

[15] Kuhn directly disqualifies this kind of "thought experiment" in the first manuscript when he says that in fact we cannot even be certain that Einstein could convince Newton (Kuhn (n.d.), M1, TSK Archives – MC240, box 4, folder 3, (7)). But he does not do this yet in the Lowell Lectures (1951) (see Mayoral 2017, 264, on Kuhn and Conant). On the social character of science, see Wray 2011. In the Thalheimer Lectures (1984), edited by Pablo Melogno in Spanish (Kuhn 1984b/2017), Kuhn seems particularly clear on this subject. See also Melogno 2019.

PART II

Three Core Concepts

CHAPTER 5

Mop-Up Work
William Goodwin

5.1 Introduction

Thomas Kuhn conceived of himself, in the project most notably realized in SSR, as presenting and supporting a novel image of science grounded in both its history and practice. Kuhn offered his image as an alternative to what he took to be a previously existing image of science, one that he characterized in various ways at various times, but which can be usefully summarized as the image of science natural to a philosophically inclined student of physics in early post–Second World War America. Kuhn was so inclined and had been so trained, and thus presumably he shared this prior image, at least to some extent, up until receiving his PhD in physics and beginning his sustained encounters with the history and philosophy of science. These encounters, occasioned by his invitation to teach in James Conant's course on the history of the experimental sciences and then facilitated by his membership in the Harvard Society of Fellows, where he read broadly in both history and philosophy, eventually resulted in a transformation of Kuhn's personal image of science. SSR, along with some significant surrounding texts, is fruitfully conceived of as an attempt to recreate in his audience Kuhn's own personal transformation to a new image of science.

Central to Kuhn's new image of science is the idea that there are two distinct modes of scientific inquiry, its normal mode and its extraordinary mode. These modes show up along several dimensions of Kuhn's image, including, but not limited to, the historical development of science, the cognitive strategies appropriate for its success, its social organization, and in the "location" and nature of its content. Often, Kuhn frames the differences between his image of science and any that he hopes to oppose or subvert as resulting from the neglect of one or the other of these modes, as they manifest along one of these particular dimensions. For instance, according to Kuhn, the mistaken understanding of scientific development as cumulative comes from exclusive focus on normal science and the consequent neglect of the noncumulative changes occasioned by extraordinary science. The misleading ideology of the scientist as creative freethinker, on the other hand, comes from exclusive focus on extraordinary science and neglect of the sort of rigid reasoning required by the normal mode.[1] Thus, these two distinct modes are often Kuhn's primary analytic tools for articulating what is importantly different about his image of science, and they are typically characterized in opposition to one another along some particular dimension. One cannot, therefore, hope to understand normal science without sustained encounters with extraordinary science as well. Furthermore, full appreciation of Kuhn's notion of normal science requires understanding how these dual modes manifest themselves as important differences along each of the several dimensions of Kuhn's image. Because Kuhn is a systematic thinker, the ways that these modes show up in one dimension constrain or reinforce the way they show up in all of the others. This means that, though one may start by appreciating how these modes are characterized in opposition to one another in one dimension of Kuhn's thought − say in the historical development of the basic sciences − one will not have a full appreciation of the strength of Kuhn's vision until one sees how this duality brings the entirety of Kuhn's image, in all of its dimensions, into focus. My effort to capture and convey Kuhn's notion of normal science will begin with three metaphors that he uses to introduce and explain the significance of the normal mode in his thinking about science. I will then introduce a metaphor of my own for understanding Kuhn's significance, particularly through the notion of normal science, for contemporary work in the philosophy of science.

5.2 Scientist as Custodian

Kuhn reports[2] that the notion of normal science first enters his thought in a paper called "The Function of Measurement in Modern Physical Science" (reprinted in Kuhn 1977). This paper contrasts a "textbook" image of the roles of measurement in physical science with the roles of

---

[1] A prime example of this occurs in "Logic of Discovery or Psychology of Research" (reprinted in ET) where Kuhn accuses Popper of mistaking extraordinary science for all of science thereby leading to a misleading account of scientific development.
[2] From (RSS, 295): "Just that little phrase very early on about an extended mopping-up operation − I don't even remember quite how it gets introduced, but that's where the notion of normal science enters my thinking."

measurement that Kuhn finds in the history and practice of the physical sciences. Not surprisingly, given what was claimed earlier, central to Kuhn's alternative image of measurement is the distinction between "normal" and "extraordinary" measurement. Normal measurement is, of course, the sort of measurement that takes place during normal science, which is itself characterized in this paper as follows: "The bulk of scientific practice is thus a complex and consuming mopping-up operation that consolidates the ground made available by the most recent theoretical breakthrough and that provides essential preparation for the breakthrough to follow. In such mopping-up operations, measurement has its overwhelmingly most common scientific function." (ET, 168) As this passage makes clear, Kuhn's image of measurement is bound up with a broader vision of how science progresses – through theoretical breakthroughs followed by periods of consolidation. Not only does this quote, therefore, provide a useful entry point for beginning to understand Kuhn's notion of normal science, but it also demonstrates the interdependence of the various dimensions of Kuhn's image of science. I will begin my attempt to articulate the role of normal science in Kuhn's thought by highlighting, and explicating in turn, two central features of normal science evident in this quote. First is the descriptive historical claim that the overwhelming majority of scientific work takes place in the normal mode and that this mode is punctuated, periodically, by "breakthroughs" or episodes of extraordinary science. Second is a functional claim about the respective roles of normal science and these periodic breakthroughs in scientific development.

Kuhn makes something like this descriptive claim about the prevalence of normal science in almost every text where he introduces the term. For instance, later in this article, Kuhn describes normal science as, "the sort of practice in which all scientists are mostly, and most scientists are always, engaged" (ET, 177). Not only does the typical scientist in a mature, basic science never have the opportunity to engage in extraordinary science, but even those that do have such opportunities still spend the majority of their careers engaged in normal science. Kuhn explains that opportunities for extraordinary science are rare and depend on the overall state of the scientific community, not solely on the dispositions of the individual scientist. Individuals within a mature scientific community cannot simply choose to become extraordinary scientists; they are instead trained to be normal scientists, and in most cases both their success as scientists and the success of science – that is, progress within their scientific field – depend upon their continuing, along with the bulk of their peers, to work in that normal mode. Eventually, however, one of the reliable products of all that work in normal science is the creation of opportunities for extraordinary science through the erosion of the consensus of the normal science community. Extraordinary science, in turn, often regenerates fresh opportunities for normal science by providing the material around which a new consensus can form. Scientific development, in other words, is characterized by extended periods of normal science punctuated by rare episodes of extraordinary science, each essential to the other. And each mode contributes to scientific progress in its own distinctive way.

Kuhn's descriptive characterization of scientific development amounts to the idea that a certain pattern is evident in the history of the sciences. He supports this idea by locating this pattern in the development of several specific scientific disciplines and then claiming that these disciplines are historically typical. Because evaluating Kuhn's historical generalization both exceeds my professional capacities and would take us too far afield, I will instead focus on clarifying what Kuhn claims to find repeated throughout the history of the basic sciences. There are two related, qualitative differences between disciplinary communities that are central to Kuhn's historical pattern. The first difference is in the structure of the community. Some communities – frequently, but not exclusively, those that we view as mature sciences – show evidence of having reached a consensus about the foundation of their discipline. Within such communities, there is seldom "overt disagreement over fundamentals" and instead there is a shared sense of how to go on within the field. Evidence for such a consensus can be found in the emergence of societies or professional organizations devoted to such an approach, as well as in distinctive forms of communication and pedagogy. For instance, esoteric journals become the preferred mode of communication when fundamentals can be taken for granted, and more dogmatic approaches to pedagogy, textbooks as opposed to historical texts, become appropriate ways to bring initiates into the fold. The second qualitative difference between disciplinary communities is in the nature of their work. Some communities support a "puzzle-solving tradition" while others do not.[3] Kuhn's puzzle-solving metaphor is the subject of the next section, and so here I will just bring out the principle sorts of historical evidence to which Kuhn points in order to establish the existence of such a tradition. Progress, Kuhn holds, is the

---

[3] Kuhn becomes clearer about the distinction between sharing a consensus and supporting a puzzle-solving tradition in several of his post-SSR works, particularly those that engage with Popper and his "demarcation condition." See in particular RSS, 138–140, which describes the conditions that a discipline must satisfy in order to support a "puzzle-solving tradition." See also ET, 274–77.

surest sign of a disciplinary community that has settled on a puzzle-solving tradition. Progress shows up as a steady accumulation of increasingly articulated and specialized solved problems. Within a puzzle-solving tradition, there are expectations about what sorts of problems should be solvable and what counts as a solved problem, and so progress is particularly evident within communities that have established such a tradition. Furthermore, within such traditions failure to solve a puzzle is typically attributed to the researcher, not the tradition itself. Kuhn acknowledges that other sorts of communities might show evidence of progress; representational painting is one of his examples, even though they do not support a puzzle-solving tradition. But the kind of progress evident in the mature sciences such as modern physics is found only in disciplines that support puzzle-solving.

With these distinctions in hand, it is possible to describe the large-scale pattern that Kuhn finds evident in the history of the sciences. What we now regard as mature, basic sciences take place in disciplinary communities that both share a consensus about fundamentals and support a puzzle-solving tradition. However, as one traces the development of these fields backward in time, one finds that there are periodic shifts in the puzzle-solving traditions, and other aspects of the consensus, governing the field. These earlier traditions are at the very least incompatible, and perhaps even incommensurable, with the subsequent approach in the field. The famous "revolutions" in the history of science, such as the Chemical Revolution or the Quantum Revolution, mark the transitions between traditions that Kuhn has in mind here. These shifts between puzzle-solving traditions take place in relatively short, localized segments of disciplinary development that have distinctive characteristics. These characteristics are the historical signs of extraordinary science and they include the blurring of the consensus within the discipline as well as "the willingness to try anything, the expression of explicit discontent, the recourse to philosophy and to debate over fundamentals" (SSR-4, 91). During extraordinary science, in other words, a discipline shares a problem-solving tradition, albeit, one that is in crisis, but lacks a consensus about how to go on. Or, it might be better to say, they have a diminished consensus. Sometimes, the history of a discipline will show evidence of several such transitions between incompatible problem-solving traditions, each marked by a breakdown in the consensus of the community. However, eventually, each discipline can be traced back to an original moment when it first reached maturity, that is, when it reached a consensus that was strong enough to support a problem-solving tradition. Before this original moment, disciplines are in what

Kuhn refers to as a pre-paradigm state; they are not mature sciences and instead should be regarded as "proto-sciences."[4] In its pre-paradigm state, what will become a mature, scientific discipline is often fragmented into distinct schools, each of which is governed by a distinct consensus about how to approach the field. However, none of these schools is yet able to support, or establish, a problem-solving tradition. Because of this, none of the proto-sciences make the distinctive sort of progress characteristic of science, though they may all contribute, in one way or another, to the eventual emergence of a mature science. This emergence occurs only when one or another of the schools eventually develops a consensus that is sufficiently rich that it supports a puzzle-solving tradition and thereby wins over the bulk of the community working in that discipline.

On Kuhn's view, scientific work in the normal mode is not, as Popper suggests, "a danger to science and, indeed, to our civilization" (Popper 1970, 53); instead, it has crucial functions in scientific development. There are two principal ways that normal science contributes to this development, one intentional on the part of participating normal scientists and one inadvertent. As described in the quote that begins this section, the first of these is to consolidate "the ground made available by the latest theoretical breakthrough." Consolidating the latest theoretical breakthrough means – and this is the main insight of "The Function of Measurement in Modern Physical Science" – generating precise expectations about how the relevant parts of the world should behave based on that theoretical viewpoint, often, as Kuhn argues in this paper, by engaging in "normal" measurement and establishing norms of "reasonable agreement." For the participating normal scientists, this can involve not only refining the expectations in those domains where their theoretical viewpoint has already had success, but also extrapolating that approach into novel domains and thus generating expectations about new parts of the world. When the normal scientist manages to bring the world into alignment with her expectations, we have the solved problems that are characteristic of scientific progress. Furthermore, while engaging in these sorts of problem-solving projects, normal scientists are also, inadvertently, contributing to the development of science in another way. They are preparing for the "breakthrough to follow" by creating opportunities for the world to fail to meet the increasingly precise expectations they generate from the current theoretical viewpoint. As Kuhn puts it, sometimes normal science pays "an additional dividend: something may go wrong" (ET, 171). Opportunities

---

[4] See RSS, 138, for the introduction and characterization of "proto-science."

for things to go wrong are created both because more precise expectations make it possible to see – and harder to explain away – discrepancies between expectation and measurement and also because new parts of the world are explored in the quest to extrapolate the current theoretical approach. When things go wrong – and we should expect that they will, according to Kuhn – opportunities for extraordinary science are created.[5]

Normal scientists, which as we have seen means most scientists, most of the time, are custodians of the approach to science that they work within: they refine the approach, defend it from anomalies, and push it into new corners of the world. Their work plays crucial roles in scientific development, both intentional and inadvertent. Furthermore, Kuhn often emphasizes the difficulty, persistence, innovation, and excitement that such work can involve, and so it is not because the work of normal science is mundane that Kuhn refers to it as mop-up work. Rather it is because normal science takes place in the context of a fixed approach to science – the approach of which those normal scientists are custodians – that Kuhn describes it in these terms. Though in the early work described in this section Kuhn characterizes the approach to science within which a normal scientist works only in terms of the "theoretical development," which leads to it, eventually, Kuhn will broaden his conception of what is involved in an "approach to science" within which a normal scientist works. It is to Kuhn's more developed accounts of what is involved in solving problems in accord with an approach to normal science that we must now turn.

5.3 Scientist as Puzzle-Solver

The second metaphor that is central to Kuhn's thinking about normal science equates the work of the normal scientist with puzzle-solving. Generating and vindicating expectations about how the world will behave according to the latest theoretical development "requires the solution of all sorts of complex instrumental, conceptual, and mathematical puzzles" (SSR-4, 36). This equation is meant to emphasize several features of the work of normal science. These features not only figure prominently in Kuhn's attempt to characterize normal science, but they are also used both to reveal tensions in the image of science that Kuhn hopes to overthrow

---

[5] At least for the large scale "theoretical innovations" of the sort that seem to be at stake in "The Function of Measurement in Modern Physical Science," anomalies themselves are not, however, sufficient to precipitate a crisis, thereby initiating extraordinary science (see SSR-4, chapter 7). Smaller scale changes to normal science can be initiated, at least in some cases, simply when things go wrong (see SSR-4, chapter 6).

and to motivate alternatives to that image. The first feature that the puzzle metaphor is supposed to bring out about normal science is that it is "traditional" work requiring "convergent" thinking: "normal research, even the best of it, is a highly convergent activity based firmly upon a settled consensus acquired from scientific education and reinforced by subsequent life in the profession" (ET, 227). This means that there are constraints – very much like the rules governing puzzle solutions – within which the normal scientist must work. Second, normal science does not aspire to overturn the traditions on which it is based; instead, the "most striking feature" of the work of normal scientists is "how little they aim to produce major novelties, conceptual or phenomenal" (SSR-4, 35). Just as when at work on a chess problem, the rules of the game are presupposed in normal science and are not, generally, "corrigible by normal science at all" (SSR-4, 122). I will first expand upon the sense in which normal science is both governed by rules and presupposes – rather than attempts to undermine – a tradition and then address the ways that puzzle-solving is supposed to both reveal flaws in the standard image of science and suggest alternatives to it.

The basic contention about normal science that grounds the thought that it is governed by rules is the existence of a "strong network of commitments – conceptual, theoretical, instrumental, and methodological" shared by the practitioners within a scientific tradition (SSR-4, 42). These shared commitments not only determine what sorts of objects and relations the world contains, but also the "methods, problem-field, and standards of solution" appropriate to investigating that world (SSR-4, 103). In the context of such shared commitments, normal scientists agree not only about the basic structure of the world they are investigating, but also about what questions are worth addressing, what would constitute satisfying answers to these questions, and what methods and instruments are appropriate to such investigations. With these things in place, puzzle-solving and thus scientific progress become possible. It is important to Kuhn, at least by the time that he gets to SSR, that the rules of the normal science game are understood to include more than just "theory." He emphasizes that the shared commitments of scientists can be both considerably more abstract, and considerably more concrete, than statements of scientific laws or principles. More abstractly, scientists often share "quasi-metaphysical commitments" that restrict them to puzzle-solutions consistent with one or another metaphysical understanding of nature. For instance, in an example Kuhn often uses, early theories of electrical attraction were typically mechanistic, postulating material effluvia emanating from charged bodies rather than spooky, Newtonian action-at-a-distance because the scientists in question were committed to the idea that all interaction was by contact (see Roller and Roller 1957). On the more concrete side, scientists typically share substantial procedural or instrumental commitments, which constrain how and whether measurements bear on the puzzles they are trying to solve. For instance, few psychological studies these days would be regarded as significant if they were not blinded in appropriate ways. Collectively, then, these rules act as constraints – at multiple different levels – on the practice of normal science, and within these constraints, scientists are able to decide both what sorts of problems should be solvable and when they have been successfully solved.

Another way that Kuhn attempts to characterize normal science is by cataloging the sorts of problems upon which normal scientists work. This catalog is useful because it provides examples not only of how scientists are constrained by rules, but also of the senses in which they take their research tradition for granted in order to engage in that work, and thus are not aiming to undermine that tradition when engaging in this work. Roughly speaking, Kuhn contends that there are three types of work in which the normal scientist can be engaged.[6] These are (i) fact gathering, (ii) enhancing the contact between theoretical approach and the world, and (iii) articulation of the approach. A classic example of fact gathering would be the work of Berzelius who "devoted the greater part of 10 years of his life to the careful determination of some 2000 combining weights of elements and compounds" (Nash 1957, 301). Though perhaps not particularly interesting to outsiders, this careful laboratory work was crucial to the advance of the atomic theory because it allowed, eventually, for accurate determination of atomic weights and molecular formulas. This is why Berzelius engaged in it. He was not trying to overthrow or undermine the atomic theory, but rather to supply the crucial facts that would solidify that theoretical approach.

Ray Davis, a 2002 Nobel Prize winner in Physics, provides a wonderful example of the next sort of problem for normal science – finding ways to bring theory and world into contact (Davis Jr., 2019). Davis worked for thirty years developing a solar neutrino detector. Neutrinos are notoriously difficult to detect and so considerable ingenuity was required not only to come up with some situation in which solar neutrinos might be detected,

---

[6] Kuhn is clear that these kinds of problems are not neatly resolvable; any particular problem in normal science may involve elements of all three kinds. Additionally, he distinguishes between laboratory and theoretical work – also admitting that these are not cleanly separable – but I will treat these together in what follows.

and to develop expectations for what should be detected in those circumstances, but also to create and implement a detection apparatus. The results from Davis's neutrino detector led eventually to the Solar Neutrino anomaly because he did not detect the expected number of neutrinos coming from the Sun. Just as Kuhn might have predicted, this anomaly was initially taken to indicate that either Davis's detector was flawed or that the expectations were wrong, and not that there was something amiss in our understanding of neutrinos. Only after independent corroboration was Davis's result understood to be correct and consistent with expectation, and our theory of neutrinos revised to explain the initial discrepancy. What is relevant about this story, besides its demonstration of the elaborate work that normal scientists engage in in order to set up a potential point of contact between theory and measurement, is that Davis's work took the theory of neutrinos for granted. Davis tried to bring the theory of neutrinos into contact with theories of solar fusion – his experiment was premised on substantial assumptions about how neutrinos worked – even though, in the end, his work has led to substantial changes in our understanding of neutrinos. It was an unintentional side benefit, and not the intention, of Davis's thirty-year effort to count radioactive argon atoms that substantial changes were made in our understanding of neutrinos.

The last category of normal scientific work is articulation, which is the "most important" class of problems in normal science (SSR-4, 27). Articulation is especially important because it aspires to increase the precision of a tradition's expectations and to push those expectations into new territory. Having increasingly precise expectations about an ever-larger domain is what creates the opportunities for the tradition to fail and, eventually, transform into extraordinary science. Returning to the Solar Neutrino anomaly will allow for a good example of articulation as well. John Bahcall was the astrophysicist who calculated the expected flux of solar neutrinos that Ray Davis had set out to measure (Tremaine 2011). Combining theories of solar fusion with the physics of the neutrino, he was able to develop, also over the course of thirty years, increasingly precise predictions for how many radioactive argon atoms Davis should be counting in his detector. This work made neutrino physics relevant to solar physics in a new way thereby creating novel opportunities for the world to live up to our expectations. When the world failed to do so, normal scientists, working to maintain and extend the traditions in which they were trained, nonetheless created opportunities for substantial, noncumulative changes to those traditions.

According to Kuhn, scientific education and subsequent practice prepares the scientist for puzzle-solving within a normal scientific tradition and so this "rigorous training in convergent thought" supplies skills that are crucial to scientific progress. The fact that scientists are trained in this way explains why "preconception and resistance [to change] seem the rule rather than the exception in mature scientific development" (Kuhn 1963, 302). At the same time, however, "scientists are taught to regard themselves as explorer[s] and inventors who know no rules except those dictated by nature itself" (Kuhn 1963, 313). Instead of as a tradition-bound custodian toiling to maintain an inherited approach, the scientist learns to think of himself as a "man who rejects prejudice at the threshold of his laboratory, who collects and examines the bare and objective facts, and whose allegiance is to such facts and them alone" (Kuhn 1963, 301). If Kuhn is right about the prevalence of normal science, then this part of the professional ideology of science is not, however, reflected in the working life of most scientists, most of the time. This reveals a tension between what scientists actually do and how they regard themselves, at least insofar as they inherit their self-image from the prevalent image of science that Kuhn hopes to undermine.

Obviously, Kuhn hopes that revealing this tension between scientific practice and the standard image of science will make his own image more appealing by contrast. However, the ideology of the scientist as the divergent, free thinker does play some important roles in our understanding of science. It explains both the appeal of the life of the scientist, to boldly go where no man has gone before, as well as how science has repeatedly managed to radically transform our understanding of the world; freethinking is not constrained by the current traditions, and so it should not be surprising that freethinkers regularly upset them. While Kuhn is prepared to admit that science as a collective enterprise "does from time to time prove useful, open up new territory, display order, and test long accepted belief," he maintains that "*the individual* engaged on a normal research problem *is almost never doing any one of these things*" (SSR-4, 38). He therefore needs alternative accounts both of the appeal of the scientific life and of how a community of rule-bound traditionalists end up generating radically new understandings of the world.

The puzzle-solving metaphor is important to both of these accounts. First, puzzle-solving is supposed to provide an alternative explanation of what makes the work of normal science satisfying. What challenges the normal scientist, according to Kuhn, "is the conviction that, if only he is skillful enough, he will succeed in solving a puzzle that no one before has solved or solved so well" (SSR-4, 38). And second, it is at the community level that the satisfying, but tradition-bound work of normal science generates the sorts of novelty characteristic of science's self-image. As we saw in the examples mentioned earlier, dedicated puzzle-solvers – whether those puzzles involve counting argon, decomposing and weighing chemical compounds, or predicting solar neutrino flux – can contribute in essential ways to solidifying or initiating substantial changes in our scientific understanding. When puzzles are solved by bringing the world into alignment with our expectations, those solved puzzles can reinforce an emerging consensus. When puzzles instead cannot be solved or ignored, eventually the community moves into extraordinary mode where divergent thinking is tolerated and sometimes required. Typically, it is not the well-entrenched puzzle-solvers who have revealed the crisis-causing anomalies that go on to supply the ingredients of the new consensus. Instead it is outsiders, or younger scientists less bound to the tradition, and liberated by an emerging crisis, who are able to suggest modifications of the community's "rules" that bring anomaly back into line with expectation. Individual diversity in the context of shared commitments thus helps explain how scientific communities largely composed of intricate puzzle-solvers can nonetheless be "most noted for the persistent production of novel ideas and techniques" (ET, 232).

5.4 Confining the World in Very Special Boxes

In addition to the custodial and puzzle-solving metaphors introduced in the previous sections, Kuhn was also fond of characterizing normal science as an attempt to force nature into boxes. He says, for instance, "closely examined, whether historically or in the contemporary laboratory, [normal science] seems an attempt to force nature in the preformed and relatively inflexible box that the paradigm supplies" (SSR-4, 24).[7] I take it that the major point of characterizing normal science this way is to emphasize that the custodial work of the individual normal scientist is constrained by the approach to science within which she works, and for the most part, those constraints are subject to neither her discretion nor her modification. Instead, a scientific approach, or a "paradigm" as Kuhn refers to it in this quote, is community property. Being a custodian to a paradigm means

---

[7] Kuhn (RSS, 159) asserts: "[T]he history of proto-science shows that normal science is possible only with very special boxes, and the history of developed science shows that nature will not indefinitely be confined to any which scientists have constructed so far."

acknowledging, or at least working in accord with, a set of community commitments. These community commitments are the constraints, or boxes, into which the work of the normal scientist must fit. Because the existence of normal science is what "most nearly distinguishes science from other enterprises" (ET, 272) and normal science requires a community consensus that supports a puzzle-solving tradition, science for Kuhn is an essentially social enterprise. Without a likeminded community of fellow puzzle-solvers, a scientist cannot be engaged in science at all.[8] However, as noted in the last section, Kuhn also is committed to the idea that individual diversity among the members of a scientific community plays crucial roles in scientific development. The boxes into which a normal scientific community must force nature must be very special indeed; they must be firm enough to support a puzzle-solving tradition, but flexible enough to allow for the diversity crucial to scientific advance. This section will therefore present Kuhn's account of the nature of the group commitments that support puzzle-solving, as well as his explanation of how these commitments are consistent with the essential diversity of scientists within the group so constrained. Doing so will also require appreciating Kuhn's distinctive account of the "location" and nature of the cognitive content of science and how that account interacts with Kuhn's essentially social conception of science in order to explain scientific development.

It must be admitted (as indeed he does, see SSR-4, 181) that Kuhn's attempts to characterize the shared commitments of communities of normal scientists took many different forms in SSR, not to mention in surrounding work. These attempts show up as the various characterizations of the paradigms that guide or constrain a normal science tradition. Masterman famously distinguished twenty-one different senses of "paradigm" in SSR (Masterman 1970, 61), but for our purposes it will suffice to distinguish two. In some places, Kuhn equates a paradigm with a shared set of "conceptual boxes" including, but not limited to, a worldview (SSR-4, 5), while in others he characterizes paradigms much more concretely as particular achievements upon which the ongoing work of normal science is modeled (SSR-4, 11). The former is a "paradigm" in the broad sense, while the latter is a "paradigm" in the narrow sense. A paradigm in the broad sense is a "constellation of group commitments" constraining scientific work (SSR-4, 181). Such commitments must amount to, metaphorically

---

[8] Addressing a particular example of a proto-science, Kuhn claims: "[T]hough the field's practitioners were scientists, the net result of their activity was something less than science" (SSR-4, 13). He means, I think, that the field did not show evidence of the sort of progress characteristic of normal science, which is, for him, an essentially social phenomenon.

speaking, a very special set of boxes that supports both a puzzle-solving tradition and the essential social diversity of science. A paradigm in the narrow sense is a concrete model for future scientific work and amounts to the most distinctive feature of Kuhn's account of scientific content. Fortunately, at least partially in response to Masterman's analysis, Kuhn devoted substantial attention after SSR to refining his accounts of paradigms in both the broad and narrow senses. Better still, he introduced new terminology that eliminated the unfortunate equivocation infecting his most prominent terminological contribution to contemporary academic life.

In the "Postscript" to SSR, Kuhn introduces a new term for the constellation of commitments shared by a normal science community or what was referred to earlier as a paradigm in the broad sense. He calls such a set of commitments a "disciplinary matrix." The term "matrix" is meant to emphasize that these group commitments have a specific and ordered structure; not any old boxes will do. The sorts of commitments that make up the components of the disciplinary matrix are what make the distinctive sorts of progress characteristic of mature basic science possible. To put this another way, Kuhn maintains that it is because its practitioners share commitments of the sort spelled out in the disciplinary matrix that normal science supports a puzzle-solving tradition that is also flexible enough to allow for innovation at the community level.

Though Kuhn does not offer a comprehensive list of the components of a disciplinary matrix, he does offer a partial list and then distinguishes one particular component as centrally important. That centrally important component is what was referred to earlier as a paradigm in the narrow sense. The disciplinary matrix[9] includes symbolic generalizations, models, values, and what he now refers to as "exemplars." Symbolic generalizations are, Kuhn maintains, the vehicles by which mathematics and logic are imported into normal scientific work. They include the laws and abstractly stated principles that make up the theories traditionally invoked by philosophers of science and they allow for, among other things, quantitative measurement, which is one primary way of making our expectations more precise. Models, on the other hand, encompass more abstract constraints

---

[9] This account is drawn from the Postscript to SSR. "Second Thoughts on Paradigms" contains another, slightly different, characterization of the disciplinary matrix and exemplars (ET, 297–319). Additionally, there are some important commitments mentioned in the original text of SSR, but not included in these characterizations of the disciplinary matrix. Most obvious among these are what Kuhn calls "instrumental commitments" to various types of laboratory apparatus and procedures, as well as their interpretation (see SSR-4, 40–41 and 59–61).

on a normal science tradition. These include both quasi-metaphysical visions of what the world is like and how it can be explained, as well as heuristic analogies that guide the development of a particular field (e.g., a gas as elastic billiard balls in random motion). They contribute to a puzzle-solving tradition by both setting the problem field for a discipline, suggesting possible solutions and then helping to establish standards of solution; furthermore, they create room for anomalies because only against some such background can the world fail to live up to our expectations. The values to which a scientific community is committed include, according to Kuhn, both a general regard for problem-solving and the traditional epistemic virtues of a good scientific theory, such as enhanced scope, simplicity, and accuracy (see ET, 321–2). Though also more abstract than a commitment to a theory, values constrain what sort of work the community finds acceptable, by privileging, for example, precise quantitative agreement over mere explanatory consistency. Furthermore, values often prove crucial to the choice between problem-solving traditions that becomes prominent during extraordinary science (see ET, 320–39). Exemplars, which are the components of the disciplinary matrix that correspond to paradigms in the narrow sense, are shared sets of "recurrent and quasi-standard illustrations of various theories in their conceptual, observational, and instrumental applications" (SSR-4, 43). As such, commitments to shared exemplars are much more concrete constraints on the scientific practice than a commitment to a particular theory. Concrete problem solutions serve as models for future practice, and as such they determine how it is appropriate to go on within that scientific tradition. I will discuss these in more detail in what follows. Collectively, then, the components of a disciplinary matrix support normal science's generation of precise expectations about the world, as well as the capacity to recognize when those expectations have not been met (i.e., they support a puzzle-solving tradition). This allows a community sharing a disciplinary matrix to perform the two roles crucial for the normal mode in scientific development: the generation of solved problems and the eventual recognition of significant anomalies.

It is relatively easy to see how the components of the disciplinary matrix that are more abstract than theories or symbolic generalizations are compatible with the social diversity essential to Kuhn's account of scientific development. A quasi-metaphysical view of nature is generally vague enough that many different proposed problem solutions, or even puzzle-solving traditions, are compatible with it. For instance, many different accounts of protein function, or the structure of DNA, are consistent with a commitment to a non-Vitalist approach to biology. While such a commitment does not dictate a particular approach to these concrete questions, it does suffice to rule out many others. Similarly, shared values can continue to play an important role in how a community evolves, even though "two men deeply committed to the same values may nevertheless, in particular situations, make different choices as, in fact, they do" (ET, 331). Two scientists might, for instance, both value simplicity and accuracy but differ either in how they interpret or weigh these values in a concrete situation. They might then pursue different articulations of their shared research tradition, though they agree on the values guiding the enterprise. This sort of diversity within a shared consensus is thus "the community's way of distributing risk and assuring the long-term success of its enterprise" (SSR-4, 186). Similarly, this same diversity in the interpretation of values is one way of explaining why some community members, in the midst of extraordinary science, dig in their heels and support the old paradigm, while others abandon ship and pursue much less well-developed but novel alternatives. All of the flexibility within a community that shares values and models would be irrelevant, however, if the symbolic generalizations and concrete exemplars in the disciplinary matrix were sufficient to dictate the approach of a research tradition going forward. It should not be surprising to find, therefore, that according to Kuhn's account of scientific content, this is not the case.[10]

Fundamental to Kuhn's account of scientific content is his recognition that linguistic formulations of scientific laws, theories, or concepts are essentially incomplete. That is, to whatever extent scientists within a normal science tradition can agree on such linguistic formulations, those formulations do not, by themselves, capture all the cognitive constraints on future work in that tradition. In SSR, Kuhn presents this as a discoverable historical fact (it is very difficult to come up with consensus formulations, but consensus models are easy to identify), which is also made evident by the pedagogical strategies of science education. As Kuhn (ET, 287–8) puts it in contrasting his views with Popper: "The books and teachers from whom [scientific knowledge] is acquired present concrete examples together with a multitude of theoretical generalizations. Both are essential carriers of knowledge." Philosophical accounts of science that limit scientific content to the linguistic formulations of theories are thus in conflict with both the history and pedagogical practice of science.

---

[10] For an elaboration of the sorts of disagreements to be expected on Kuhn's account, see Goodwin (2013).

Instead, Kuhn contends that in most cases, actually applying a scientific law or theory to a concrete situation requires learning to recognize that concrete situation as analogous to a previously accepted model solution. These previously accepted model solutions are what one works through during the course of scientific training, first in science textbooks, and then in the journal literature. Without having worked through model solutions, abstract formulations of theories or laws by themselves do not specify how they are to be attached to nature or related to the outputs of instrumentation. Concrete models, or exemplars, are crucial intermediaries between theories, or more specifically, abstract generalizations, and the world.[11] "Theory" is not one of Kuhn's terms, but if it were, and it meant something like the total cognitive content of a scientific tradition, then Kuhn might well endorse the claim that a scientific theory includes a collection of models, where those models are particular, exemplary puzzle solutions. It is because of this view that Kuhn regards his puzzle-solving analogy as fundamentally flawed. Whereas one can precisely explicate in the case of a typical puzzle the rules for acceptable solutions, this is not possible in normal science. Instead, there is an essential role for tacit knowledge, learning to recognize analogies or similarities that cannot be fully articulated, but which also need not be fully articulated in order to explain the ongoing coherence of that normal scientific tradition.

For Kuhn, the coherence of a research tradition rests, at least in part, on family resemblances with, and analogies to, the shared corpus of established achievements, the exemplars within the disciplinary matrix. When a community is constrained by analogies to a shared set of model applications, however, there is an irreducible open-endedness about how those analogies will be extended into novel territory. Thus, Kuhn's views on scientific content are not limited to the insufficiency of linguistic formulations but extend to the insight that linguistic entities of this sort are also essentially open-ended. That is, for Kuhn, the application of these laws, theories or concepts is not fixed for every conceivable situation to which they might apply. Unlike a logical theory that specifies in advance the extensions of its predicates and thus the scope of its theoretical claims, scientists must work with criteria that "are sufficient to answer that question only for the cases that clearly do fit or that are clearly irrelevant" (ET, 288). The most basic reason for this is that, according to Kuhn, scientific

---

[11] Barnes (2004) argues that Kuhn's theory of content is fundamentally a recognition of the "intransitivity of empirical sameness" and thereby brings out another important sense in which science is essentially social.

concepts are "legislative" (ET, 258–60); that is, our use of a particular concept, or set of concepts, presupposes, often implicitly, certain expectations of nature. Our expectations of nature within some normal scientific tradition are, in turn, developed not by learning logically complete characterizations of the world, but instead by working through shared exemplars. These shared exemplars demonstrate clear cases where the relevant concepts apply and show how to integrate those concepts with the "symbolic generalizations" of the field in those particular circumstances. But knowing how to apply, or whether a concept applies, in clear cases that model an approach to science, does not suggest that there are clear criteria for the application of scientific concepts in the unexpected circumstances that normal science – according to Kuhn – reliably generates. This is why, as Kuhn puts it, "confronted with the unexpected, [the scientist] must always do more research in order further to articulate his theory in the area that has just become problematic" (ET, 288). So our concepts, and thus our symbolic generalizations and experimental procedures, encode our expectations of the world. Because those expectations are not logically codified in advance but are rather extrapolations from concrete problem solutions, the various practitioners of a normal science tradition can, at least occasionally, diverge in how they propose to go on. Confronting a novel situation, those practitioners may disagree about how or whether a particular concept applies. This may result in different strategies for integrating that novel phenomenon into the shared scientific tradition, based perhaps on the idiosyncratic psychological or historical features of the different practitioners.[12] So the indeterminacy of scientific content, which issues from its being partially encoded in concrete exemplars, can be seen to contribute to the constrained diversity crucial for Kuhn's social epistemological account of scientific development. I hope then to have established that Kuhn's very special boxes make it plausible that a scientific community is both sufficiently constrained to support a puzzle-solving tradition and sufficiently open-ended to allow for the required social diversity.

---

[12] Compare this with Mark Wilson's reading of Kuhn in *Wandering Significance* (Wilson 2006, 282–3). While Wilson congratulates Kuhn for recognizing the fundamental indeterminacy of our scientific concepts, and the role of personal factors in deciding how individual scientists will go on, he accuses Kuhn of appealing to paradigms to fix the content of our concepts. I am suggesting that even within the same paradigm, the content of a concept is not fixed for Kuhn, particularly for novel cases, and that this is crucial to his account of scientific development.

5.5 Science in Two-Point Perspective

Science, for Kuhn, is a distinctive epistemic phenomenon – it progresses in a way unlike other human enterprises. Over time, science produces ever more articulated and precise "knowledge" of our world. This "knowledge" is manifest in the array of problems – intellectual, instrumental, and practical – that science allows us to solve. It was a pressing philosophical concern for Kuhn to explain this important and unusual human enterprise – to understand what made this distinct form of progress possible. Kuhn's approach to this question was shaped by his own background. Trained as a basic scientist, and then as a historian, he wanted a philosophical account that was compatible with both the historical facts and the practice of science as he understood them. This put him in conflict with the most prominent philosophical accounts of scientific development on offer at the time. When Kuhn began reflecting on these issues, mainstream philosophy of science emphasized individual inquirers, modeled on the famous innovators of science: Newton, Darwin, Einstein. These individuals were understood to follow logical rules, either inductive logic or conjectures and refutations, thereby producing novel theories – conceived in analogy to theories in mathematical logic – which constituted and added to settled scientific knowledge. The resulting image of science was incompatible with Kuhn's experience as a practicing scientist and historian. Many of these conflicts have been described earlier in this essay, as has been Kuhn's basic response: to insist that science can only be satisfactorily rendered in two-point perspective. As it is a messy human enterprise, any philosophical account of science will flatten its object, but, Kuhn found, capturing its important dimensions – those crucial to understanding how it progresses – requires two ideal points, not just one. Representing science with one vanishing point – be it normal science or extraordinary science – obscures the details necessary to appreciate its distinctive developmental pattern. On Kuhn's approach, the two modes of science, though they are still idealized, allow us to see several central features of the enterprise that are crucial to understanding how it progresses, but which had been systematically overlooked by the mainstream philosophy of science of his time. Many of these features are now standard and central concerns of the philosophy of science. Thus "normal science" and its essential opposite "extraordinary science" have had a profound impact on the philosophy of science whether or not they are themselves, as analytic tools, too idealized or otherwise insufficient to provide a philosophical rendering of science.

Central among the enduring features of philosophical concern brought out by Kuhn's two-point perspective on science are its essential social structure and the extent to which the cognitive content of science must extend beyond traditional, logical theories. As we have seen, scientific development, for Kuhn, could only be understood in terms of a diverse community of inquirers working under shared constraints, which vary with the mode that science is in. What constraints are in play vary by the mode that science is in. These inquirers have their own distinctive histories and personalities, which result in different contributions to the development of their shared discipline. The constraints that unite them encompass many different aspects of scientific practice and are not limited to endorsing particular theories. As has become increasingly common in the philosophy of science, Kuhn drew attention not just to the theories that scientists use to characterize the world, but also to the instruments and experimental procedures that need, somehow, to be brought to bear on those abstract cognitive tools. Likewise, he saw an essential role not only for the values of scientists, particularly when it comes to reestablishing normal science after a period of extraordinary science, but also for their more abstract and metaphysical perspectives. Most presciently, perhaps, from the point of view of contemporary philosophy of science, was his identification of the central importance of models in scientific inquiry. Not only does the disciplinary matrix include a prominent place for models, in Hesse's sense (see Hesse 1963), but also Kuhn's appeal to exemplars, or concrete problem solutions, is a recognition that much of the time scientists work in a context where, as Masterman puts it, "the theory is not there" (1970, 66). Going on, as a scientist, Kuhn saw, is much more about seeing analogies with concrete problem solutions that have worked in the past than it is about unpacking the consequences of previously established theories. This insight has been usefully developed throughout contemporary philosophy of science (see e.g., Giere 2004, and Cartwright 1983). I hope to have shown that "normal science" is one half of an analytic pair that has, in Kuhn's hands, allowed for a much more realistic and multidimensional rendering of our scientific attempts to come to terms with the world.

CHAPTER 6

Kuhn and the Varieties of Incommensurability
William J. Devlin

6.1 Introduction

One of the integral, but highly controversial, concepts in Thomas Kuhn's philosophy of science is his thesis of incommensurability. Introduced in SSR (1962), Kuhn maintains that members of competing scientific paradigms are inevitably unable to make complete contact with another as, to some extent, there is no common measure between them. In this chapter, I examine the varieties of Kuhn's notion of incommensurability. I analyze its broader understanding as found in SSR, which gives rise to three forms of incommensurability: *methodological*, *semantic*, and *observational*. However, Kuhn later takes a "linguistic turn" to emphasize the possibility and role of translation implicit in a paradigm shift. His development of incommensurability concludes with a modified semantic version known as *local incommensurability*, where lexicons are incommensurable when it is impossible to translate from a local subgroup of terms and sentences into another subgroup.

I contend that, through the varieties of incommensurability, Kuhn establishes reasons to criticize and reject any form of the correspondence theory of truth that is predicated upon a mind-independent world. First, I demonstrate that under observational incommensurability and the lexicon framework of local incommensurability, Kuhn offers an argument that challenges the epistemic assumptions behind truth as correspondence. Second, I suggest that underlying both semantic and local incommensurability, Kuhn assumes Merrill B. and Jaakko Hintikka's conception of language as a universal medium. Given this conception of language, along with Kuhn's rejection of a neutral language, I argue that Kuhn provides reasons to challenge the linguistic assumptions behind the correspondence theory.

6.2 Three Periods of Incommensurability

Kuhn explains that he borrowed the term "incommensurability" from mathematics, where, for instance, the hypotenuse of an isosceles right triangle is incommensurable with its side insofar as there is no common measure expressible in whole numbers (see RSS, 189 and 35). He found that this absence of a common measure was analogous to his experiences interpreting old scientific texts. Initially, Kuhn considered obsolete scientific terms of old theories to be nonsensical until he realized they were being misread: "[T]he appearance of nonsense could be removed by recovering older meanings for some of the terms involved, meanings different from those subsequently current" (RSS, 91). That is, successive scientific theories were incommensurable, as they lacked a common measure between them. Both Kuhn and Paul Feyerabend introduced incommensurability separately in publications in 1962: Kuhn in SSR; Feyerabend in "Explanation, Reduction, and Empiricism." Kuhn recounts that his own use of the term was broader. While Feyerabend restricted incommensurability specifically to the language between two theories, Kuhn opened incommensurability to further include differences in methods and standards between theories (RSS, 33–34, fn. 2).

Elsewhere, Kuhn relates that he found this thesis to be the most important idea presented in SSR: "No other aspect of SSR has concerned me so deeply in the thirty years since the book was written" (RSS, 91). He continued to develop his thesis over the next three decades after SSR. We can identify at least four different forms of incommensurability across its historical development. In this section, I highlight Kuhn's varieties of incommensurability across three periods of Kuhn's publications.[1] Kuhn's robust thesis of incommensurability in SSR – what I call the "early period" – includes three forms of incommensurability: (1) *methodological incommensurability*, (2) *semantic incommensurability*, and (3) *observational incommensurability*. During the aftermath of SSR in the 1970s – the "middle period" – Kuhn builds upon semantic incommensurability, drawing on W. V. O. Quine's indeterminacy of translation. Lastly during the 1980s and 1990s – the "later period" – Kuhn introduces a final version of incommensurability he calls *local incommensurability*.

---

[1] Howard Sankey traces Kuhn's versions of incommensurability in this three-phase manner (Sankey, 1993). I follow this approach to not only delineate the varieties of incommensurability, but also elucidate the aspects of incommensurability relevant to Kuhn's criticism of the correspondence theory of truth. For a similar historical overview in relation to prominent interpretations and criticisms, see Dirk-Martin Grube (2013); for an assessment of Kuhn's historical and evolutionary developments of incommensurability in relation to both reality and truth, see James A. Marcum (2015a).

6.2.1 The Early Period: The Original Thesis of Incommensurability in SSR

In SSR, incommensurability is embedded in the notions of paradigms and revolutions. A scientific paradigm, or disciplinary matrix, eventually faces a crisis, which entails a scientific revolution, or "those non-cumulative developmental episodes in which an older paradigm is replaced in whole or in part by an incompatible new one" (SSR-4, 92). Kuhn later elaborates on this incompatibility, introducing his thesis of incommensurability: "[P]roponents of competing paradigms must fail to make complete contact with each other's viewpoints" as they "are always at least slightly at cross-purposes" (SSR-4, 147). Kuhn expounds on "cross-purposes" by offering three components, or forms of incommensurability, that constitute the essence of his original thesis.

First, there is a *methodological incommensurability* between pre- and post-revolutionary paradigms, as each paradigm is its own "source of the methods, problem-field, and standards of solution accepted by any mature scientific community" (SSR-4, 103). There is an incommensurability of standards or definitions of science, where "proponents of competing paradigms ... often disagree about the list of problems that any candidate must resolve. Their standards or their definitions of science are not the same" (SSR-4, 147). For instance, proponents of Aristotelian science assume that an explanation of the cause of attractive forces between particles of matter is a necessary condition for a theory of motion. Meanwhile, Newton's dynamics implies it is not necessary; rather, it is sufficient to only note the existence of such forces. Proponents of the competing "Aristotelian" and "Newtonian" paradigms thus disagree about whether or not the explanation of the cause of attractive forces is a problem that must be resolved.

As Kuhn explains in his 1969 "Postscript" to SSR, this standard variance is exacerbated since proponents of competing paradigms cannot find a common neutral "ground" upon which to stand to determine a list of problems as a standard or definition of science: "[T]here is no neutral algorithm of theory-choice, no systematic decision procedure which, properly applied, must lead each individual in the group to the same decision" (SSR-4, 198). The determination of standards or definitions of science is internal to the paradigm. Without a neutral algorithm, paradigms remain incommensurable methodologically.

Second, Kuhn maintains that paradigms are incommensurable semantically, as there is a conceptual disparity between the vocabularies embedded in competing paradigms. Kuhn describes this *semantic incommensurability* as follows: "Within the new paradigm, old terms, concepts, and experiments fall into new relationships one with the other. The inevitable result is ... a misunderstanding between the two competing schools" (SSR-4, 148). Though the post-revolutionary paradigm may borrow terminology from the pre-revolutionary paradigm, the meaning of these terms changes. As Kuhn explains, "since the vocabularies in which they discuss such situations consist ... of the same terms, they must be attaching some of those terms to nature differently" (SSR-4, 197). Two competing paradigms have a different network of relations between concepts, resulting in different ways for how a concept is used or attached to nature. For example, "the transition to Einstein's universe" required that "the whole conceptual web" of terms (space, force, etc.) "be shifted and laid down again on nature whole" (SSR-4, 148). As such, communication across paradigms is only partial.

Third, Kuhn explains that a fundamental aspect of incommensurability is *observational incommensurability*. Here, Kuhn offers such controversial expressions as "when paradigms change, the world itself changes with them" (SSR-4, 111) and "the proponents of competing paradigms practice their trades in different worlds" (SSR-4, 149). To elaborate on "different worlds," Kuhn utilizes theory-dependence of observation, developed from N. R. Hanson's (1958) thesis that all observations are theory-laden. Here, perception is context-dependent in that the context of background theories and beliefs have an influence upon our perceptions so that what one sees "depends both upon what he looks at and also upon what his previous visual-conceptual experiences has taught him" (SSR-4, 113). Drawing upon analogies to gestalt psychology, Kuhn holds that just as our descriptions of a gestalt image are influenced by our background beliefs, so too our empirical observations are informed by the paradigm under which we practice.

To draw out the significance of observational incommensurability, let us distinguish between two forms of theory-dependence: *strong* and *weak* forms. The strong form holds that observation is a cognitive achievement as background beliefs influence the process of observation; that is, they influence background beliefs so strongly that it determines our perception that something is, or is not, the case. Two observers with different background beliefs see two different things when observing the same event (see Churchland, 1979, 1988). Meanwhile, the weak form holds that background beliefs do not influence our perceptions; instead, they influence the inferences we make, or our process of deriving judgments from the observations we have, about the world. In this case, two observers with different background beliefs categorize their observations differently, thereby attributing a different relevance to the same event – they describe and interpret phenomena in different ways (see Fodor, 1984, 1988).

In developing observational incommensurability, Kuhn suggests a strong form of theory-dependence, using the example of a swinging stone. Here, the Aristotelian sees constrained fall: the Galilean, a pendulum motion. For Kuhn, some may suggest a weak form of theory-dependence, as background paradigms may determine one's interpretation of an event. That is, the Aristotelian and Galilean "both saw pendulums, but they differed in their interpretations of what they had seen" (SSR-4, 120). But Kuhn rejects this account, instead favoring the strong form. The Aristotelian and Galilean "see different things when looking at the same sort of objects ... the first saw constrained fall, the second a pendulum" (SSR-4, 120–121). Under observational incommensurability, the process of the scientist transitioning from one paradigm to the other "is not one that resembles interpretation" (SSR-4, 121). Instead, it is like "wearing inverting lenses" so that one confronts the same event finding it now significantly "transformed" (SSR-4, 121).

This transformation suggests, metaphysically, the world is paradigm-dependent. But I agree with Paul Hoyningen-Huene, who identifies two meanings of "world" in SSR. First, there is the mind-independent world or the world as it exists independent of human consciousness. Second, there is the dependent world or the world whose content (at least partially) depends on the human mind – including one's background paradigm. Kuhn maintains that one lives in the world of experience as a "world already perceptually and conceptually subdivided in a certain way" (SSR-4, 129). At the same time, the dependent world is co-determined by the mind-independent world: it is "determined jointly by nature [i.e., the mind-independent world] and paradigms" (SSR-4, 125).

Furthermore, Kuhn adopts, what Hoyningen-Huene calls the plurality-of-phenomenal-worlds thesis (1993, 36), the thesis that there are multiple dependent worlds, some merely possible and some actualized, based on the acceptance and employment of competing paradigms. The world in which the scientist observes, experiences, and practices is uniquely determined by the mind-independent world and the paradigm employed by the scientific community. After a paradigm change, the world the scientist experiences is properly understood as a different dependent world, a world uniquely determined by the mind-independent world and the new paradigm.

6.2.2 The Middle Period: Semantic Incommensurability and Translation

Though paradigms and revolutions may have initially played more prominent roles in SSR, incommensurability had perhaps the largest impact, drawing both support and criticism. In the 1970s, Kuhn defends his thesis from criticisms by primarily developing semantic incommensurability. Continuing to emphasize the analogy to its mathematical counterpart, he now interprets the absence of a common measure as the loss of a neutral language. Kuhn holds that, should two successive theories be semantically commensurable, there would be a "point-by-point comparison" between them for a successful communication (RSS, 162). To allow for a full communication between competing theories, Kuhn maintains a common or neutral language is required to help serve as a measuring scale for evaluation. This hypothetical neutral language would consist of "pure sense-datum terms" or a vocabulary that attaches to nature "in ways that are unproblematic" and "independent of" any scientific theory about nature (RSS, 162). Such a theory-independent language would allow us to use its vocabulary, which purely and directly attaches to nature (via expression), as the common ground upon which both theories can stand. The problem for Kuhn is that we do not have such a neutral language (SSR-4,125–126; RSS, 164–165, 189–190, 35).[2] Without this language, two theories that contain terms with different semantic ascriptions attach these terms to nature differently and so cannot be translated into the other theory without residue or loss of meaning. Thus, the theories remain semantically incommensurable.[3]

Kuhn appeals to Quine's indeterminacy of translation thesis to demonstrate both why we cannot have recourse to a neutral language and why such a language is necessary to have a complete translation. This thesis holds that "manuals for translating one language into another can be set up in divergent ways, all compatible with the totality of speech dispositions, yet incompatible with one another" (Quine 1964, 27). Quine illustrates this thesis with the example of a linguist investigating the language of an unknown people. With no recourse to even partial linguistic

---

[2] This lack of a neutral language as a measuring scale does not mean that we lack comparability. See D. Davidson (1974) for the charge that incommensurability leads to incomparability; see Kuhn (RSS 176–195 and 196–207) for his response to this charge.

[3] It is important to note that Kuhn does not merely argue that we *currently* lack a theory-neutral language to serve as the common ground for competing or successive theories, thereby allowing for the possible future introduction of such a language that will permit us to overcome the problem of semantic incommensurability. On the contrary, Kuhn makes the stronger claim that a neutral language consisting of observation terms that attach to nature without influence from a theory is impossible. He makes it clear that while philosophers over the past several centuries have either sought out or assumed such a neutral language (SSR-4, 125–126; RSS, 162–165), he appears to endorse the abandonment of this project (RSS, 162), arguing that such an endeavor is "hopeless" (SSR-4, 126) given that "there can be no scientifically or empirically neutral system of language or concepts" (145). For a more detailed account of semantic incommensurability in relation to the loss of a neutral language as a common measure, see Alexander Bird (2000), chapter 5.

communication, the linguist is a "radical translator" as he translates the unknown language into his known language beginning with present events conspicuous to the linguist. But translation is underdetermined under such conditions. On the occasions where a rabbit scurries by and the native says, "Gavagai," the linguist confronts a variety of tentative translations: "undetached rabbit-part," "rabbit occurrence," and so on. Linguists may compose multiple translation manuals using different translations for "Gavagai." Each translation may accurately predict what the natives will say, and yet those manuals would be inconsistent with other manuals.

For Kuhn, the indeterminacy thesis helps to show that "languages cut up the world in different ways, and we have no access to a neutral sublinguistic means of reporting" (Kuhn RSS, 164).[4] The lack of a neutral language – whether it is neutral with respect to languages for all theories or to the languages of two theories being compared – shows that we cannot have a direct translation of the meaning of terms from one theory to another without a residue or loss of meaning. As Kuhn argues, theory-change involves subtle changes to the meaning or application of the theory's vocabulary. While some terms, such as "force" and "mass," may be used across theories, how they "attach to nature has somehow changed" (RSS, 162–163). Without access to a neutral language during scientific revolutions, we are not in an epistemic position to successfully translate the complete linguistic meaning of the pre-revolutionary theory into the post-revolutionary theory. Thus, scientific theories remain semantically incommensurable.

6.2.3 *The Later Period: Local Incommensurability*

In the 1980s and 1990s, Kuhn introduces a "modest" version of incommensurability he refers to as local incommensurability:

> The claim that two theories are incommensurable then is the claim that there is no language, neutral or otherwise, into which both theories, conceived as sets of sentences, can be translated without residue or loss.... Most

---

[4] While Kuhn seemed to fully embrace Quine's indeterminacy of translation as informing semantic incommensurability during the start of the middle period, he comes to distance incommensurability from this thesis. For instance, in demonstrating that incommensurability does not entail incomparability, he acknowledges that translation between two theories is possible, where translation involves compromise (see RSS, 189–190). Similarly, he recognizes that, in his account of semantic incommensurability in SSR, he conceived of a neutral language as "one in which any theory at all might be described." By the end of the middle period, however, Kuhn holds that "comparison requires only a language neutral with respect to two theories at issue" (189 fn. 20).

of the terms common to the two theories function the same way in both; their meanings ... are preserved; their translation is simply homophonic. Only for a small subgroup of (usually interdefined) terms and for sentences containing them do problems of translatability arise. The claim that two theories are incommensurable is more modest than many of its critics have supposed. (RSS, 36)

Local incommensurability is a sharpened version of his earlier semantic incommensurability, which focuses on "the conceptual vocabulary deployed in and around scientific theory" (RSS, 36). Two theories are again incompatible due to a conceptual disparity between two theories. As before, the problem concerns meaning change without a common language to assist with translation. But Kuhn's final installment of incommensurability develops his first semantic thesis by introducing at least four additional components: (1) a distinction between translation and interpretation, (2) localized incompatibility of kind terms, (3) lexicons, and (4) synchronic applications.

First, Kuhn still holds that "incommensurability ... equals untranslatability" (RSS, 60). Nevertheless, he is wary that this mantra is a "regularly overinterpreted assumption" leading to misguided critiques (RSS, 34–35). To alleviate such misunderstandings, he expands on translation, dividing "actual translation" into two processes: translation and interpretation. Translation is the process where "the translator systematically substitutes words or strings of words in the other language for words or strings of words in the text ... to produce an equivalent text in the other language" (RSS, 38). Meanwhile, interpretation is the process where one works on a text consisting of "unintelligible noises or inscriptions" (RSS, 38). The interpreter observes the behavior and contextual circumstances pertaining to the production of the text and creates hypotheses that make the noises and inscriptions intelligible. If successful, then the interpreter has learned a new language (or an earlier version of the interpreter's own language). But the success of learning a new language does not necessarily entail that the interpreter has been able to translate the new language into one's own.

Given these processes, Kuhn recasts Quine's "radical translator" as an interpreter in his experiences of observing the natives say "Gavagai." And it is at the stage of interpretation that we find the start of local incommensurability. The interpreter begins by learning to identify what elicits the term "Gavagai" from the natives and so then acquires the ability to properly use the native's term. This process of interpretation is completely independent of translation (in the narrow sense). The process of translation occurs after the interpreter has completed her interpretation.

Kuhn maintains that the interpreter may be able to successfully translate the term "Gavagai" into his own language, say English. This success does not simply entail introducing the term "Gavagai" into English; rather, it involves describing in English the referents of the term "Gavagai," such as "furry," "long-eared," and so on. The translation of "Gavagai" into English as "furry, long-eared, creatures" is successful if this is the translation one seeks. If so, then "Gavagai" can be introduced as an abbreviation of the translation. In this case, the issue of incommensurability does not arise (RSS, 39).

However, translation may not always be successful. We may find that there is no English description that is co-referential with the term "Gavagai." The failure of translation is due to what the interpreter does during the process of interpretation. Kuhn holds that "in learning to recognize gavagais, the interpreter may have learned to recognize distinguishing features unknown to English speakers and for which English supplies no descriptive terminology" (RSS, 40). The native speakers may structure the world of animals differently than the way English speakers structure the world. If so, we cannot have a direct translation where the result is co-referential terms in the native and English languages. Hence, "Gavagai" becomes a case of local incommensurability as it belongs to a subgroup of terms that cannot be directly translated, without residue or loss of meaning of the original term.

Kuhn's restriction that incommensurability applies only to a sub-class of terms brings us to a second important attribute of local incommensurability, namely that it concerns an incompatibility of taxonomic or kind terms. Kind terms – which include at least natural and artefactual kinds – are limited by what Kuhn calls the no-overlap principle: "no kind terms ... may overlap in their referents unless they are related as species to genus" (RSS, 92). For example, in our language community, the kind terms "dogs" and "cats" are mutually exclusive: If "members of a language community encounter a dog that's also a cat ... they cannot just enrich the set of category terms" as this would violate the no-overlap principle (RSS, 92). Instead, Kuhn holds that they must restructure their taxonomy. In this sense, Kuhn maintains that "incommensurability thus becomes a sort of untranslatability, localized to one or another area in which two lexical taxonomies differ" (RSS, 93).

Third, Kuhn elaborates on this untranslatability of kind terms – as stemming from the process of translation (in the narrow sense) and of interpretation – by introducing the concept of lexicons or lexical taxonomies or lexical networks. Kuhn defines a lexicon as "a conceptual scheme, where the very notion of a conceptual scheme is not that of a set of beliefs but of a particular operating mode of a mental module prerequisite to having beliefs, a mode that at once supplies and bounds the set of beliefs it is possible to conceive" (RSS, 94). Kuhn is making two points here. First, he tells us what a lexicon is not – it is not simply a set of beliefs that one may have about the world. Second, he tells us what a lexicon is – it is an operating mode of cognition that determines how one perceives and understands the world. How one comes to understand the world, then, partially depends upon the cognitive and conceptual constructions of organization that are projected upon the world. Similarly, the lexical structure "mirrors aspects of the structure of the world which the lexicon can be used to describe, and it simultaneously limits the phenomena that can be described with the lexicon's aid" (RSS, 52). Kuhn contends that lexical structures are co-determined biologically and socially, where the former is established as "products of shared phylogeny," while the latter is a result of education and "the process of socialization" (RSS, 101). Kuhn refers to his theory of lexicons as a "post-Darwinian Kantianism" insofar as both lexicons and Kant's categories supply "preconditions of possible experiences" (RSS, 104). At the same time, lexicons suggest a stronger relativism, as they "can and do change, both with time and the passage from one community to another" (RSS, 104).

Lexicons determine and constrain not only our experiences of, but also our communication about, the structure of the world (RSS, 101). In this sense, a lexicon is pre-linguistic as it "must be in place before description of the world can begin" (RSS, 92). The lexicon provides the criteria that one uses in identifying the referents designated in one's language. Through these specified criteria, the speaker applies a concept to nature by determining whether or not that concept holds for a given object, event, or activity in nature. Such lexicon-laden criteria determining concept-use is also used to help build the structure of relations between that concept and other concepts within the vocabulary of the lexicon: "Those criteria will tie some terms together and distance them from others, thus building a multidimensional structure within the lexicon" (RSS, 51). Kuhn calls this the lexical structure (RSS 196–207; RSS 90–104). Members of the same language community, who are members of the same culture (scientific or other), have the same lexical structure and so operate under a lexicon that uses shared "taxonomic categories of the world and ... similarity/difference relationships between them" (RSS, 52). The criteria used to determine concept-use does not need to be identical: Different members of a language community may apply different criteria for determining concept-use, but this does not necessitate that these speakers use a concept differently. A concept will be used the same so long as the speakers maintain the same relations between the concept used and that concept's relations to other concepts in the vocabulary: "[T]heir taxonomic structures must match, for where structure is different, the world is different, language is private, and communication ceases until one party acquires the language of the other" (RSS, 52).

Lexicons and lexical structures are central to local incommensurability. Different lexical structures "impose different structures on the world" (RSS, 52). Successful translation between two language communities does not depend on using identical terms; rather successful translation occurs only if both communities share the same lexical structure:

> The referring expressions of one language must be matchable to coreferential expressions in the other, and the lexical structures employed by the speakers of the languages must be the same, not only within each language but also from one language to the other. Taxonomy must, in short, be preserved to provide both shared categories and shared relationships between them. Where it is not, translation is impossible. (RSS, 52–53)

Should two language communities have contradictory lexicons, there are statements and theories using kind terms developed within one lexicon that could not be made or used in another (RSS, 93). For instance, though both the Copernican and Ptolemaic lexicon include the kind term, "planet," their taxonomies establish different meaning relations for the term as different kinds. The term used in one lexicon cannot consistently map onto the term used in the other without violating the no-overlap principle. While the statement "Planets travel around the Sun" can be expressed in the Copernican lexicon, it cannot be so in a meaningful way in the Ptolemaic lexicon. Thus, due to the "fundamental change in some taxonomic categories," we have an instance of local incommensurability (RSS, 94).

Finally, Kuhn further develops local incommensurability so that it pertains to scientific development in two ways. As in his previous iterations of incommensurability, this incompatibility holds diachronically within a single science in terms of the relationship between an older and more recent theory/paradigm/belief concerning the same range of phenomena (such as the Copernican and Ptolemaic paradigms). But Kuhn extends local incommensurability by cutting "a synchronic slice across the sciences rather than a diachronic slice containing one of them" (RSS, 97) so that incommensurability holds between contemporary scientific specialties. Kuhn explains this incommensurability between specializations with the analogy to speciation. Just as a unit of speciation consists of members collectively making up the gene pool, ensuring both self-perpetuation and isolation, so too, the unit of scientific specialization consists of members collectively sharing "a lexicon that provides the basis for both the conduct and the evaluation of their research ... and maintains their isolation from practitioners of other specialties" (RSS, 98). Thus each specialty has a distinct lexicon where the differences remain local.

6.3 Incommensurability and Truth

In RSS, Kuhn outlined his next projected book, PW (see Hoyningen-Huene, 2015). Along with the issue of rationality, Kuhn noted two central components to the book: (1) incommensurability and (2) "its relationship to questions of relativism, truth, and realism" (RSS, 91). Clearly, Kuhn considered incommensurability to be a challenge to traditional theories of truth and scientific realism. In this section, I turn to incommensurability and its relationship to truth, as we find that Kuhn viewed his thesis as offering reasons to reject the correspondence theory of truth (CTT). Briefly, CTT generally consists of two metaphysical propositions: P_1: Truth is a property of statements and P_2: A statement is true iff it corresponds to facts that obtain in the mind-independent world. Additionally, for the purposes of understanding Kuhn's critiques, we can introduce two corollary claims, which make CTT an operational theory of scientific inquiry. First, since truth involves correspondence to the mind-independent world, it is important to assume access to this world: C_1: We have access to the mind-independent world insofar as our world of experience is caused by, and faithfully represents, the mind-independent world. Second, CTT requires that language is the proper medium for expressing the content of the mind-independent world: C_2: Language has the capacity to represent the facts that obtain in the mind-independent world.

Kuhn consistently rejects CTT: "[W]hat is fundamentally at stake is rather the correspondence theory of truth, the notion that the goal, when evaluating scientific laws or theories, is to determine whether or not they correspond to an external, mind-independent world. It is that notion, whether in an absolute or probabilistic form, that I'm persuaded must vanish" (RSS, 95; also SSR-4, 205; RSS, 160; RSS, 114–115; RSS, 243, 245).

But while Kuhn rejects CTT altogether, I suggest that his thesis of incommensurability challenges its corollary claims, C_1 and C_2. I show that observational incommensurability and the lexicons central to local incommensurability entail a rejection of C_1; meanwhile, semantic and local incommensurability pose an epistemic problem for C_2.[5] By challenging its epistemic and linguistic assumptions, CTT becomes, practically speaking, useless for scientific inquiry and theory choice, thereby motivating Kuhn to call for an alternative account of truth.

6.3.1 Incommensurability, World Change, and Moving Archimedean Platforms

While Kuhn planned to demonstrate how incommensurability poses problems for both realism and CTT in PW, these challenges were initiated in SSR. In his "Postscript," Kuhn explains that he rejects the idea that successive theories are converging toward the truth. This view derives from the assumption that there is a match between the entities specified in the theory and what exists in the mind-independent world. But given paradigms and revolutions, Kuhn rejects this ontological assumption: "There is I think, no theory-independent way to reconstruct phrases like 'really there'; the notion of a match between the ontology of a theory and its 'real' counterpart in nature now seems to me illusive in principle" (SSR-4, 205).

Kuhn's dismissal of a theory-independent way to reconstruct such phrases is due, in part, to incommensurability. In SSR, Kuhn's central challenge to CTT is that C_1 fails, given observational incommensurability. Given theory-dependence and the plurality of dependent worlds, some of which are actualized with paradigm shifts, such incommensurability prevents accessing the mind-independent world directly, without paradigm or theory-dependence. If we had such access, then observational incommensurability would not arise, since one could use the mind-independent world as the common measure, as it were, to show that one paradigm is closer to the truth and so should be favored in theory-choice debates. Thus, observational incommensurability rejects direct epistemic access to the mind-independent world. Likewise, Kuhn's emphasis that paradigms are co-constitutive of the mind-dependent world suggests that our world of

---

[5] While methodological incommensurability poses epistemic challenges to theory choice, I do not believe it presents a specific challenge to CTT. For a detailed analysis of Kuhn's methodological incommensurability, his later introduction of epistemic values, and the idea that this incommensurability entails epistemic relativism, see Sankey (2012).

experience is not a faithful representation of the mind-independent world. At the very least, we could not separate the paradigm elements from the mind-independent world elements in the constitution of the dependent world. Observational incommensurability thus suggests a rejection of C_I.

Kuhn continues this line of criticism of C_I in local incommensurability. There, Kuhn's post-Darwinian Kantianism maintains that lexical categories precondition possible experiences and thereby, along with the mind-independent world, constitute the dependent world.[6] As Kuhn explains, "each lexicon makes possible a corresponding form of life within which the truth or falsity of propositions may be both claimed and rationally justified" (RSS, 244). But the rational justification of propositions is pragmatic insofar as they are lexicon-dependent. For instance, in the Aristotelian lexicon, propositions concerning "force" may have a truth-value whose justification is made, internal to the lexicon. Meanwhile, this truth-value has no bearing on similar propositions about "force" stipulated in the Newtonian lexicon.

This lexical-dependence of truth-values suggests how Kuhn "relativizes the [Kantian] categories (and the experienced world with them) to time, place, and culture" (RSS, 245). Scientific practice acquires a body of knowledge of the actualized "form of life the lexicon permits" (RSS, 245). A lexical change entails alternative forms of life which constitutes a different body of knowledge. Given such changes and differences, Kuhn holds that "scientific knowledge claims are necessarily evaluated from a moving, historically situated, Archimedean platform" (RSS, 96). The "moving Archimedean platform" is an epistemic vantage point that "moves with time and changes with community and sub-community, with culture, and sub-culture" (RSS, 113). With this kind of platform, the justification of propositions and bodies of knowledge remains rational, as shared beliefs help to inform evaluations of theory-change and theory-choice. But it does not produce justifications or access to truths that incorporate "absolute fixity for all times" (RSS, 245).

---

[6] In his later work, Kuhn refers to "the variety of niches within which the practitioners of ... various specialties practice their trade." While more space should be devoted to this topic, I see Kuhn as using "niche" to replace the dependent world, partially because of his concern that dependent world is confused with idealism or robust constructivism and partially because local incommensurability now includes synchronic incompatibilities. Thus, rather than refer to micro-dependent worlds encapsulating different contemporary scientific specialties, niches serve as a substitute. Regardless, like the paradigm- or lexicon-dependent world, Kuhn considers a niche to "both create and [be] created by the conceptual and instrumental tools with which their inhabitants practice upon them." At the same time, it remains "as solid, real, resistant to arbitrary change as the external [independent] world was once said to be" (RSS, 120).

In order to have access to such absolute truths – truths that correspond to the mind-independent world – Kuhn argues we need "the Archimedean platform outside of history, outside of time and space" (RSS, 115). This fixed, unmoving platform serves as the epistemic vantage point "supplied only by neutral observations," which are both "the same for all observers" and "independent of all other beliefs and theories" (RSS, 113). But given that lexicons are relative to time, place, and culture, Kuhn contends the traditional Archimedean platform "is gone beyond recall" (RSS, 115). Thus, because the fixed, unmoving Archimedean platform is unavailable, Kuhn rejects C_1. For Kuhn, the ahistorical, atemporal, and non-spatial epistemic vantage point provided in the traditional Archimedean platform is necessary to evaluate propositions in relation to the mind-independent world. It provides, as it were, epistemic access to that world. Given that such a platform is unattainable for Kuhn, we do not have epistemic access that would allow us to determine whether any given proposition corresponds to the mind-independent world under CTT.

6.3.2 Incommensurability and Language as the Universal Medium

It is clear that Kuhn took his thesis of incommensurability to provide reasons for a rejection of CTT, specifically on the grounds that CTT's epistemic assumptions as stipulated in C_1 are unwarranted. But I argue that two forms of Kuhn's incommensurability – his semantic incommensurability and local incommensurability – provide a challenge to CTT's linguistic assumptions as stipulated in C_2. I propose that, in both forms of incommensurability concerning language, Kuhn assumes a view of language that Merrill B. and Jaakko Hintikka refer to as language as the universal medium. Given both semantic or local incommensurability and language as the universal medium, Kuhn's conceptions of paradigm- and lexicon-dependent language communities entail a rejection of C_2.

The Hintikkas present a fundamental opposition between two conceptions of language, language as the universal medium (LUM) and language as calculus (LC). LUM maintains that language can only be used if we already begin with a specified interpretation or "a given network of meaning relations obtaining between language and the world" (Hintikka and Hintikka 1986, 1). If we assume a certain network of meaning relations, then we can never say anything significant or meaningful about them because language already presupposes such relations. Thus, according to LUM, we cannot stand in a position to evaluate the connection between language and the world it attempts to describe: "[O]ne cannot as it were look at one's language from outside and describe it" (1986, 1). It follows then that semantics, the field of language concerning the relationship between language and the world, is inexpressible according to LUM. As the Hintikkas explain, proponents of LUM espouse the "thesis of the *ineffability of semantics*" (1986, 2). There are thus three important views that belong to LUM: (1) one cannot step outside language and view it from the outside, (2) one cannot discuss in language the relationships that connect language to the world, and (3) conceptual truths are inexpressible, as they belong to the study of semantics (Jaakko Hintikka 1996, 25).

Meanwhile, LC holds that "all those things are possible which a universalist thinks of as being impossible" (Hintikka 1996, 26). That is, it entails three opposing viewpoints: (1) one can step outside of language and view it from the outside, (2) one can discuss in language the relationships that connect language to the world, and (3) conceptual truths are expressible. As Martin Kusch explains, the calculist conceives of language as a tool, "as something that can be manipulated, re-interpreted, improved, changed and replaced, as a whole or at least in a large scale" (1989, 4). Considering that language is no longer conceived as a universal constant, a proponent of LC can accept meta-languages that allow them to step outside of language and evaluate it. Furthermore, they can use a meta-language to discuss semantics and express those conceptual truths bounded within semantics.

Given these opposing views on language, I argue that when Kuhn discusses incommensurability in language – whether his earlier semantic incommensurability or subsequent local incommensurability – he assumes LUM. For instance, we can see these assumptions in Kuhn's lexicons and structure of lexicons. Since the lexicon serves as an operating mode of a mental module that limits the set of beliefs its members can adopt, it determines how its members come to understand and organize sense-datum. Likewise, as the lexicon is pre-linguistic, it further determines the lexical structure so that taxonomic structures of the language community are already established. This structure, in turn, mirrors parts of the structure of the world and at the same time limits the phenomena that cannot only be discussed, but also perhaps even "recognized," through the lexicon (RSS, 52). Furthermore, as discussed, two theories are locally incommensurable due to a difference in the lexical structure assumed in the language community of each theory. Kuhn's structure of the lexicon thus seems to coincide with the Hintikkas' description of LUM, where language is used only when there is already a given network of meaning relations that obtain between language and the world. For Kuhn, this given network precisely is the lexicon.

Thus, when looking at language within a given lexicon, Kuhn assumes LUM. He considers members of a language community as enclosed within a world that is at least partially determined by the lexicon and lexical structure. To an extent, the lexicon "projects the world" for members of the language community. As the structure mirrors and limits parts of the world, members come to accept the lexicon-dependent world as the only world they can come to know through their lexical structure. As such, those members of a different language community with a different lexical structure project a different structure upon the world and so cannot fully comprehend the other language community. This is local incommensurability. There appears to be both an epistemic and linguistic gap between two language communities with different lexical structures.

These gaps stem from Kuhn's assumption of LUM. Members of a language community cannot step outside of their own language – their lexicon – so that their language can be altered to fit the language belonging to a community with a different lexical structure. If this ability to step outside of one's language were permissible (as it is assumed in language as calculus), Kuhn's theses of semantic and local incommensurability would be rejected, as it would now be possible to translate from one language to another without any residue or loss of meaning. Thus, given both semantic and local incommensurability, we can see Kuhn assumes language as the universal medium when analyzing languages within a language community determined by paradigms or lexicons.

Whether we examine semantic incommensurability and paradigms or local incommensurability and lexicons and conjoin them with LUM, we find a critique of C_2 in the correspondence theory. The lexicon determines how we use our concepts and how these concepts are related to one another, thus providing for us the lexical structure (or network of meaning relations for paradigm and lexical vocabulary). The lexicon further helps to determine how language attaches to nature, thereby "projecting the world" for its members, in a sense. Whenever we discuss the world, we are using a language determined by the lexicon. Given the lexicon-laden language, we find that when we turn to other languages with different lexicons, the projected worlds between the two language communities are different, and the problem of semantic or local incommensurability arises.

Kuhn maintains that in order to overcome the incommensurability between two scientific theories of different language communities, we ultimately need reference to a neutral language, a language that consists of an unadulterated sense-datum vocabulary, which purely and directly attaches to nature. But there is no possibility of a neutral language, since all language is lexicon-laden. There is, thus, no lexicon-free language. The role of lexicons determining language explains not only why there is no neutral language, but also why the search for a neutral language is hopeless for Kuhn: A neutral language is impossible since behind every language lies a lexicon that excludes the possibility of that language consisting of pure sense-datum vocabulary that directly attaches to nature. The lexicon that determines the language distorts language in such a way that it is not possible for it to ever be neutral.

Kuhn presents this problem in his rejection of the semantic conception of truth for theory comparison. Under this conception, Kuhn argues that the common example, "'Snow is white' is true iff snow is white," can be successfully used in theory comparison only if we assume "their proponents agree about technical equivalents of such matters of fact as whether snow is white" (RSS, 161). He contends that this assumption is unproblematic so long as it was "exclusively about objective observation of nature" (RSS, 161). However, he argues that theory comparison involves a further assumption that objective observers have the same understanding of "snow is white." That is, it assumes "that the proponents of competing theories do share a neutral language adequate to the comparison of such observation reports" (RSS, 162). Kuhn, however, rejects this assumption, thereby rejecting the semantic conception of truth.

At this point, we can distinguish two related questions around a Kuhnian criticism of C_2: (1) Is language accurately portraying the facts that obtain in the mind-independent world? and (2) Can language be used to portray the facts that obtain in the mind-independent world at all? Ultimately, Kuhn suggests an argument that we cannot know the answer to either question – we can know neither if our language currently accurately portrays the facts in the mind-independent world nor if language can be the proper medium for portraying such facts. This argument hinges on the notion of the neutral paradigm- and lexicon-free language, as it appears to be a necessary condition for knowing whether or not a paradigm- or lexicon-laden language community is properly expressing the facts in the mind-independent world. In Kuhn's terminology, the neutral language would serve as the Archimedean platform to determine the relation between language and the mind-independent world. Since we lack such a fixed epistemic platform, we cannot know the relation between language and the mind-independent world. In turn, we cannot know if our current language is representing such facts or even if language has the capacity to serve as the proper medium for expressing independent facts. C_2 thus becomes a dubious claim for Kuhn.

This critique of C_2 is an epistemic criticism, which follows a similar structure to Kuhn's criticism of C_1. In order to know whether or not the language of our community has the capacity to represent the facts that obtain in the mind-independent world, we need to have a neutral place in which to stand to evaluate whether or not our language accurately represents the facts. Kuhn's criticism of C_1 posits this neutral ground as the ahistorical, atemporal, non-spatial Archimedean platform. Given observational incommensurability or the lexicon and lexical structure framing local incommensurability, he denies the possibility of such a platform. The inaccessibility to this kind of platform entails epistemic inaccessibility to facts that obtain in the mind-independent world. Meanwhile, in his criticism of C_2, this neutral ground becomes the neutral language. Semantic and local incommensurability reveal that Kuhn rejects the possibility of such a language. The impossibility of such a language entails the epistemic inaccessibility to determine whether or not our own language is representing the facts that obtain in the mind-independent world. Likewise, given LUM – a view that I suggest Kuhn assumes about the paradigm- and lexicon-laden language communities – we cannot step outside of our language and our world. As such, it is impossible for us to be in a position to adjudicate whether any language has the general capacity to represent independent facts.

6.4 Conclusion

Across the varieties of incommensurability, we find significant challenges to CTT. In Kuhn's early formulation of observational incommensurability and later development of the lexical framework attached to local incommensurability, we find a criticism of C_1. In both cases, Kuhn rejects epistemic access to the mind-independent world. In both semantic and local incommensurability, we find a criticism of C_2, suggesting that CTT's linguistic assumption is unwarranted. Though the central metaphysical claims of CTT – P_1 and P_2 – remain, incommensurability leaves CTT unoperational, since we know neither that a given statement is true in corresponding to a fact nor that the fact is accurately represented by that statement. But rather than remain agnostic toward any reference to truth, Kuhn proposes a need for a concept of truth that introduces "minimal laws of logic" and whose "essential function" is to serve as a criterion for "acceptance and rejection of a statement or a theory in the face of evidence shared by all" (RSS, 99).

I believe that Kuhn's final reason for rejecting CTT is pragmatic. With only P_1 and P_2, CTT cannot serve as that criterion. Given Kuhn's epistemic skepticism toward C_1 and C_2, CTT has no merit for evaluating propositions in a theory, given that P_2 stipulates correspondence with facts in the mind-independent world. Kuhn suggests that, given incommensurability, a new theory of truth should be intra-theoretic and so function specifically within a lexicon (RSS, 160). Kuhn suggested alternatives to CTT – from the deflationary theory (RSS, 99), to a potential pragmatist theory, to even a revised CTT where P_2 refers to the lexical-dependent world (RSS, 77).[7] And though, unfortunately, Kuhn did not complete his outline of this theory for PW, I nevertheless agree with his view that a strong conception of truth remains "badly needed" (RSS, 99).

---

[7] Marcum adds, Kuhn's concept of truth "functions pragmatically as an instrument for eliminating lexical statements that might contradict one another" and "enhances the coherence of a community's lexicon" where communication promotes progress (2015a, 130).

CHAPTER 7

Reassessing the Notion of a Kuhnian Revolution
What Happened in Twentieth-Century Chemistry

Eric R. Scerri

7.1 Introduction

As commentators have almost universally concluded, the work of Thomas Kuhn has had a profound influence on the history and philosophy of science and numerous other fields of study. Kuhn's SSR has been subjected to a plethora of criticisms, which led him to clarify, and in many cases revise, his initial views on most of the key ideas that he famously proposed. Kuhn scholarship continues to flourish and indeed underwent a recent revival following the fiftieth anniversary of the publication of SSR (Richards and Daston, 2016; Devlin and Bokulich, 2015). It may be fair to say that there is now an equally extensive literature addressing the views of the later Kuhn as there is addressing the views of commentators on the first edition of SSR.

The present chapter seeks to explore the change in Kuhn's views on scientific revolutions in particular. I begin by citing the views of Alexander Bird who writes the following passage that I take to be of some significance for what is to follow in the present article:

> But it seems that whatever definition we employ it remains the case that the normal science versus revolutionary science dichotomy cannot do justice to the variety of episodes in science. Kuhn's terminology gives an artificial sense of there being two quite distinct kinds of scientific change. Reflection on this variety suggests that the distribution of episodes is not bimodal but instead shows a greater degree of continuity, with intermediate cases being not especially less frequent than the extremes. (Bird, 2000, 54)

My own analysis of Kuhn's understanding of scientific revolutions will be carried out by narrowing the scope of enquiry to consider the views of the editor of this volume, Brad Wray, who over recent years has developed an interest in the history and philosophy of chemistry. Wray has published a defense of Kuhn's notion of scientific revolution by proposing a possible new revolution that concerns how twentieth-century chemists changed the criterion for identifying a chemical element from atomic weight to that of atomic number.

In a recently published article, Wray claims that chemistry underwent a significant change in theory in the twentieth century and that this represents a "classic textbook case of a Kuhnian revolution" (Wray 2018a, 209). In so doing Wray refers to what he calls a "new conceptual understanding of what it is to be an element." In the present chapter I examine these claims and I revisit the episode that Wray is referring to, in order to see whether it should indeed be regarded as a classic Kuhnian revolution.

I especially want to consider (i) the notion that there was a change in theory, (ii) the idea of a classic Kuhnian revolution, a phrase that contains a good deal of ambiguity, and (iii) the question of a conceptual understanding of elements.

7.2 A Change in Theory

I believe that many scientists might be tempted to dismiss Wray's claim for a change in chemical theory on first hearing of it. Nevertheless, I also believe that his proposal is more intriguing and subtle than may appear at first. However, I find it puzzling that Wray uses the term "classic Kuhnian revolution," since this terminology would seem to invoke the original sense of how Kuhn proposed that scientific revolutions occur. While it may be correct to claim, as Wray does, that chemistry underwent profound changes regarding its underlying theory, I do not believe that the episode that Wray describes has much to do with change in theory as generally understood in science.

Wray's alleged classic revolution concerns the fact that chemists turned away from using atomic weight to order the elements in favor of the use of atomic number. However, this does not appear to be a change in theory by any stretch of the imagination. The theoretical change that was taking place at the start of the twentieth century was rather the abandonment of classical mechanics in favor of quantum mechanics.

According to the generally held account, this program began with Planck's discovery of the quantum of action in 1900,[1] followed by Bohr's application of quantum concepts to the hydrogen atom in 1913. In the same series of articles Bohr also provided a semi-empirical explanation for the

---

[1] Interestingly, according to Kuhn, Planck continued to treat energy in a continuous manner in his article of 1900.

periodic table by drawing on a judicious mixture of chemical and spectroscopic arguments in order to deduce the electronic configurations of many elements in the periodic table (Bohr 1913). This revolution in theory, if one insists on calling it so, was continued by Sommerfeld's relativistic extension of Bohr's theory and the introduction of elliptical orbits. Next came Pauli's exclusion principle, Heisenberg's uncertainty principle, and Schrödinger's wave equation. Soon afterward quantum mechanics was applied to chemical bonding by Heitler and London and eventually it became the underlying theoretical description for most aspects of chemistry.

Wray does not mention any of these far more revolutionary theoretical changes but instead locates what *he* calls a revolutionary change in theory in the transition from the use of atomic weight for ordering the elements to the use of atomic number. I would like to suggest that this development may be more akin to a change of focus from the entire atom (atomic weight) to just the number of protons (atomic number) rather than any theoretical change.

Wray also associates this change with what he interprets as a Kuhnian anomaly. Wray is referring to the fact that certain pairs of elements such as tellurium and iodine are incorrectly ordered if atomic weight is used, whereas they fall into a chemically correct sequence if one uses atomic number instead. Although I am happy to concede that such pair reversals represented anomalies in the general sense, I do not believe that they represent Kuhnian anomalies, as I will attempt to explain.

Before moving on let me just concede that Wray's proposal seems to have a certain appeal in that it features a clear-cut case of anomalies that were recognized by the chemical community and not merely labeled as such by Kuhn or his followers. This case would also seem to offer a good counter-example to Toulmin's objections that scientific discoveries often occur in the absence of anomalies (Toulmin, 1970). The case that Wray proposes is not subject to such a critique since the change in chemists' thinking was indeed motivated by what chemists themselves referred to as anomalies.

However, there are a total of only four such anomalies in the entire periodic table, even if one considers the modern periodic table that extends up to element 118. At the time that Wray is considering, only two such anomalies were known, namely those involving tellurium and iodine as well as cobalt and nickel.[2] The correlation between atomic weight and atomic number is in fact an extremely good one and as a result these anomalies were not unduly

---

[2] The other two anomalies feature the elements argon and protactinium, neither of which was known at the beginning of the twentieth century when chemists switched from using atomic weight to using atomic number to order the elements.

troublesome. Moreover, there are no known cases for which the reversal involved anything but immediately adjacent elements.

I suggest that the problem that originally existed in the two mentioned cases was nowhere as pressing as Wray seems to believe. As I. B. Cohen has written, "the profundity of a revolution in science can be gauged by the virulence of conservative attacks as by the radical changes in scientific thought that it produces" (Cohen, 1985, 414). As far as I am aware there were no virulent attacks in any part of the chemical community following the proposal that ordering of the elements should be based on atomic number instead of atomic weights and nor did it involve any radical changes in scientific thought. According to Cohen's view, at the very least, we may not therefore be dealing with a scientific revolution.

As I see it, the anomalies involving the pair reversal of certain elements do not warrant being regarded as Kuhnian anomalies that would result in a scientific revolution. It is a historical fact that many of the discoverers of the periodic system such as Newlands, Odling, Hinrichs, Lothar Meyer, and most famously Mendeleev, all recognized the need to reverse the placement of these elements, although they lacked any fundamental justification for doing so (Scerri 2020, 73–112).

Let us compare this case with what is generally taken to be a genuine Kuhnian anomaly, such as the advance of the perihelion of Mercury that necessitated a genuine Kuhnian revolution before it could be resolved. It is not so much the small differences in atomic weight connected with the pair reversals that make me doubt their possible status as Kuhnian anomalies. After all, the perihelion case also amounted to an extremely small departure from what could be calculated from the then available Newtonian mechanics. It is more the fact that pair reversals were readily tamed and did not demand any theoretical changes.

Of course, it was gratifying when the justification for pair reversal was provided as a result of the discovery of atomic number. However, neither chemists nor physicists lost much sleep over the matter during the fifty or so intervening years between the discovery of the periodic table and the change to atomic number as the more correct ordering principle.

Pair reversals certainly never amounted to a Kuhnian crisis as I see it. In any case, there was absolutely no change in the order in which any element was presented in the periodic table, either before or after the adoption of atomic number. All that happened was that the correct insight of chemists that, for example, tellurium should precede iodine, was given a physical basis following the move to using atomic number. One might say that physicists were able to catch up with what chemists already knew, solely on the grounds of chemical reactivity. These historical circumstances, and especially the complete lack of any change to the periodic table before and after the introduction of atomic number, do not bode well for those who wish to view this development as a scientific revolution.

In ending this section, we should also recall that Kuhn himself was careful to note the variety of scientific discoveries and whether they might count as examples of his way of characterizing scientific change. According to Kuhn the discovery of X-rays provoked resistance but not a crisis, and the revision was easily assimilated without any struggle between competing paradigms. As Kuhn also wrote, "if an anomaly is to evoke crisis, it must be more than just an anomaly" (SSR-1, 82). I believe that the case being considered in the present article would also be considered as rather marginal by Kuhn since it invoked neither resistance nor any form of struggle between paradigms.

The other anomaly that Wray discusses concerns the discovery of isotopes, that is to say forms of the same element having different atomic weights, which I will turn to in due course.

7.3 What Is a Classic Kuhnian Revolution?

A Kuhnian revolution in the way in which it is generally understood involves the notions of paradigm, anomalies, crises, and revolution all leading to a new paradigm. As is well known, this account met with a great deal of resistance from many critics from the time that Kuhn first published it (Scheffler 1967; Shapere 1964; Toulmin 1972). In particular Kuhn was criticized for his notion of paradigms and for claiming that the change from one paradigm to a subsequent one was associated with incommensurability or an inability of scientists from opposing paradigms to even communicate with each other.

The Kuhnian revolution that Wray proposes and describes as being "classic" would not in fact seem to be classic since it lacks most of the qualities that Kuhn assigns to scientific revolutions. I can only conclude that Wray is not referring to the writings of the "classic Kuhn" but to his later writings. Or as Wray explains Kuhn's later view, "[i]nstead of referring to theories as paradigms, [Kuhn] came to believe that theories are scientific lexicons. Each theory is a scientific vocabulary that orders the relevant concepts in specific ways, with very precise relationships between the concepts." (Wray 2018a, 210)

As is well known, in more than thirty years following the publication of his 1962 book, Kuhn constantly revised many of the central themes in his philosophical account of theory change. For example, Kuhn changed the meaning of the term paradigm in such a way that it came to mean the work of a far more restricted group of scientists than he had originally suggested. Similarly, the term "scientific revolution" changed because, in his lexical turn, Kuhn turned his attention to the language that scientists use.

Wray has been a leading contributor to the field of Kuhn scholarship and is far better acquainted than I am with the twists and turns in Kuhn's later thinking about science and how he may have revised his seemingly very "revolutionary" views set out in his book of 1962.

I therefore find it a little surprising that Wray persists in using the phrase "classic Kuhnian revolution" thus blurring the issue of whether he means the original Kuhnian account or the highly revised account following Kuhn's later turn to the lexicon of science.

My quick gloss on this issue would be to say that if the phrase is intended in the earlier Kuhnian sense, then I completely disagree with claiming that the change from atomic weight to atomic number represents such a revolution. If on the other hand Wray intends the phrase "classic Kuhnian revolution" in the later sense, in which all attention is directed at the language and terms used by proponents of rival paradigms, then he may have a stronger case, although I will have some further comments on this possible interpretation.

7.4 Brief Digression: Vogt and Kragh's Views of the Same Scientific Episode

Wray is not the only author to have suggested that the change from atomic weight to atomic number represents a Kuhn-style revolution. In a recent book review the chemist Thomas Vogt seems to believe that the change in the definition of an element that occurred as a result of the work of Van den Broek and the isotope crisis was "radical" and that it constituted a scientific revolution (Vogt 2017, 108). He then says, "The resolution of this 'isotope crisis' during the first 25 years of the 20th century had all the scientific, historical, and political complexities of a scientific revolution and is described in detail by Kragh (2000) . . . After this scientific revolution chemists never saw Nature at the microphysical level as before." (Vogt 2017, 108)

It is rather unfortunate that Vogt should have chosen this particular source since Kragh does not seem to consider the change in the understanding of what constitutes an element, and the discovery of isotopes, to have been a revolution in the Kuhnian sense.[3] As Kragh explains:

---

[3] Vogt makes no reference whatsoever to the work of the later Kuhn such that one can only suppose that he is referring to the classic and well-known earlier Kuhnian account.

> Great theoretical changes occurred during the period, but these did not lead to a wholesale refutation of older chemical concepts such as the periodic table and the notion of an element. The periodic system survived the revolution and although the chemical element was reconceptualized it occurred in such a way that continuity with the older definition was secured. (Kragh 2000, 447)

Admittedly Kragh uses the word "revolution," which might be why Vogt chose to cite him. However, it should be noted how Kragh describes this change in rather muted and very un-Kuhnian terms. According to Kragh there was no "wholesale refutation of the older chemical concepts," and the reconceptualization occurred in a way that secured "continuity with the older definition." Whatever kind of revolution Kragh might be referring to it does not seem to resemble a revolution as envisaged by the earlier Kuhn. A few lines later Kragh writes,

> Neither quantum mechanics nor the proton-neutron model of the nucleus necessitated further changes. The element and the periodic system are thus examples of conceptually robust chemical entities. Their histories indicate the force of the pragmatic chemical viewpoint and the value of retaining older theoretical notions, at least in a correspondence-like manner and up to a point. The reinterpretation of the element that occurred in the period kept the connection with the older concept through the principle of conservation of the elements in all chemical transformations. (Kragh 2000, 448)

Contrary to Vogt and more recently Wray's view of a Kuhnian style revolution in the concept of an element, Kragh emphasizes "conceptual robustness," "the value of retaining older theoretical notions" and keeping a connection with the older concept of an element. Indeed, Kragh also draws support from the work of the historian Mary Jo Nye and agrees with her that "chemistry and physics were beginning to share disciplinary terrain" (Nye 1989, 448). This too does not sound like a case of Kuhnian incommensurability associated with a scientific revolution. Indeed, the final sentence of Kragh's article emphasizes this point even further. "To Aston and many of his colleagues, there were no fundamental disagreements between physics and chemistry, only different ways of conceptualization and presentation" (Kragh 2000, 448).

However, unlike Vogt, I believe that Wray may be claiming that this episode is a revolution in the sense of Kuhn's later writings, to which I turn more fully in the following sections.

7.5 Returning to Wray's Own Proposed Revolution

Following the preliminary comments mentioned earlier, I now examine Wray's article in greater detail. It would appear that Wray is equivocating on the sense in which he is interpreting a "Kuhnian revolution." As I have already suggested, by calling the episode under discussion a "classic textbook case" Wray would seem to be referring to the original and generally well-known sense intended by Kuhn. It may also be worth noting in passing that what the later Kuhn thought about scientific revolutions has not in fact permeated science textbooks.[4]

The impression that Wray might be referring to the original sense of a Kuhnian revolution is further strengthened on hearing that the example that he proposes to discuss represents "another case." This too would seem to point to the classic cases of scientific revolutions that Kuhn discussed in his classic book of 1962.

In the opening remarks of the article that I am discussing, Wray states that

> [t]he aim of this paper is to provide an analysis of the discovery of atomic number and its effects on chemistry. The paper aims to show that this is a classic textbook case of a Kuhnian scientific revolution. The analysis serves two purposes. First, it provides another case of a Kuhnian revolution, thus offering support for Thomas Kuhn's theory of scientific change, which has been subjected to criticism on an ongoing basis. (Wray 2018a, 209)

Wray immediately follows these remarks by the statement that the change in the meaning of an element "provides a compelling unifying narrative of some of the most important research in chemistry in the early twentieth century" (Wray 2018a, 209). Although I have not conducted a poll among my colleagues in the chemistry department at UCLA, I think I can safely say that the change from using atomic weight to using atomic number, and the change in the way an element is defined, would be very unlikely to make a list of "the most important research in chemistry in the early twentieth century."

Wray explains further how Kuhn's original understanding of paradigm changed in his later writings:

> Kuhn regarded this new characterization of theory change as merely a clarification of his original account, presented in *Structure*. Even he

---

[4] Not only has the later Kuhn not entered into science textbooks, but even the earlier Kuhn is very seldom, if ever, mentioned. Scientific textbooks remain steeped in logical positivism. A casual survey of about fifty chemistry textbooks that I have conducted confirms this point. Perhaps the situation is different in physics education but I doubt it.

came to realize that the notion of a paradigm was far from clear. In his later writings, the term "paradigm" was reserved for those specific scientific accomplishments that become templates for solving other related outstanding scientific problems. (Wray 2018a, 211)

However, it would appear that the notion of a crisis, another classic term from the earlier and better-known Kuhn, is retained in his later writings. As Wray explains: "Much of the apparatus associated with the earlier paradigm-related notion of revolutionary theory change was retained in Kuhn's more recently developed lexical change model of theory change. For example, he continued to believe that anomalies played a crucial role in the process of scientific change that ultimately lead to revolutionary changes of theory." (Wray 2018a, 212)

I highlight this point because I do not think that Wray provides much evidence for the occurrence of any crises. Indeed, if anything Wray's account appears to highlight the protracted, piecemeal, and disparate contributions that led up to the change from the use of atomic weight to that of atomic number in order to identify chemical elements.

Wray proceeds to give a historical summary of nineteenth-century chemistry, especially as it pertains to the discovery of elements and the measurement of their atomic weights, leading up to the discovery of the periodic system in the 1860s.

According to Wray, who prefers to regard historical events through the lens of Kuhn, the occurrence of pair reversals represented the first important anomaly that would eventually precipitate what he nominates as a classic Kuhnian revolution. In making this claim he accepts that the early discoverers of the periodic system were perfectly capable of dealing with this anomaly. Wray writes,

> Anomalous phenomena have to be dealt with in some way, and Mendeleev and other chemists felt that such a solution was reasonable, given their knowledge of the properties of the various elements. Other pairs of elements had also posed similar problems for chemists, as long as they assumed that atomic weight was the key to classifying chemical elements, including, for example, potassium and argon, and cobalt and nickel.[5]
>
> Anomalies provide the research topics in a normal scientific tradition. And scientists sometimes choose to set some anomalies aside, to await the efforts of future scientists who may be better equipped, conceptually and technologically, to tackle the problems. Ultimately, though, these specific

---

[5] As a matter of fact, the pair reversal involving argon was not an issue for the discoverers of the periodic table for the simple reason that this element was not discovered until 1894.

anomalies did contribute to bringing about a radical change of theory in chemistry. (Wray 2018a, 213)

It would appear that Wray is claiming that pair reversals, a situation that did not constitute a crisis in chemistry, nevertheless would ultimately contribute to a revolution. Wray then makes a further claim in support of a classic Kuhnian revolution when writing:

> The result, though, was a radical change of theory of just the sort that Kuhn would regard as a revolutionary change of theory.
>   Perhaps most significant in this process was the discovery of atomic number ... The discovery follows the pattern that Kuhn identifies in *Structure*, in his analysis of the discovery of oxygen (see Kuhn 1962/2012, pp. 53–57). Consequently, it is fruitless and futile to attempt to pinpoint who discovered atomic number, and when exactly the discovery was made. (Wray 2018a, 214)

These statements may contain a contradiction as well as a factual error. If the discovery in question was indeed similar to that of oxygen, namely gradual, and not attributable to any particular event, it raises the question of whether this can be regarded as a radical change. Secondly, if ever there was a scientific discovery that could be pin-pointed rather accurately it would indeed be that of atomic number. There is little doubt among historians of science that the theoretical groundwork was laid by the Dutch econometrician and amateur scientist Anton Van den Broek, while the experimental confirmation was supplied by the English physicist Henry Moseley (Scerri 2018).[6]

Wray continues by discussing the discovery of isotopes and how they initially presented an additional anomaly to the pair reversal of certain elements. In highlighting the importance of the discovery of atomic number as providing a better means of identifying any element, Wray also explains how this development helped scientists to isolate the seven elements that remained to be discovered within the old boundaries of the periodic table.[7] In this instance I believe that he may be overstating his case.

Although focusing on atomic number and its characterization via X-ray spectra did serve to direct the search for these elements, it by no means

---

[6] Wray mentions the contribution of Van den Broek but is under the misapprehension that Van den Broek carried out some "laboratory work." In fact, all of Van den Broek's key articles were of a completely theoretical nature.

[7] I am referring to the limits that encompassed the original sequence of naturally occurring elements from H for which $Z = 1$ to U for which $Z = 92$.

settled the numerous claims that were made over a thirty or so years period, until they were all discovered or synthesized.[8] In fact, in several instances the scientists concerned claimed to provide X-ray data to support what they believed to be the isolation of one of these missing elements, only to be refuted by further research on the part of others (Scerri 2013, 183–185).

7.6 The Dual Sense of the Concept of Element

I believe that there is a missing central theme that is highly relevant to Wray's claim regarding the change in the understanding of an element as being defined by its atomic number rather than by its atomic weight. I am referring to a long-standing discussion in the philosophy of chemistry literature concerning the dual meaning of the term "element." An element can be regarded in the sense of Lavoisier as the last stage in chemical composition of a compound. This sense of element was called a simple substance by Lavoisier and was highly influential in the course of what is generally regarded as *the* Chemical Revolution that was centered around his work.

The other sense of "element" is sometimes labeled as element as a "basic substance," meaning as a fundamental substance that underlies all the properties that an element manifests as a simple substance (Scerri and Ghibaudi, 2020).[9] It is this sense of element that survives when simple substances enter into chemical combination to form compounds. Sodium chloride, to take an example, does not contain sodium and chlorine as simple substances but only as basic substances. This distinction was revived by none other than Mendeleev, the leading discoverer of the periodic table, who went as far as to say that the periodic table was primarily a classification of elements as basic substances and not simple substances.

> It is useful in this sense to make a clear distinction between the conception of an element as a separate homogeneous substance, and as a material but invisible part of a compound. Mercury oxide does not contain two simple bodies, a gas and a metal, but two elements, mercury and oxygen, which, when free, are a gas and a metal. Neither mercury as a metal nor oxygen as

---

[8] Wray claims that the rate at which these elements were discovered was quite striking. This is not the case since it required a period of about thirty years between the discovery of protactinium and the synthesis of promethium. The discovery of the seven missing elements owes as much if not more to the development of new technologies such as particle accelerators in which five of the seven elements in question were artificially synthesized.

[9] This terminology is due to the translation of Paneth's paper into English as carried out by his son Heinz Post. Some authors have objected to the use of the word "substance" in the definition of both senses of the concept of "element" (Earley, 2005).

a gas is contained in mercury oxide; it only contains the substance of the elements, just as steam only contains the substance of ice, but not ice itself, or as corn contains the substance of the seed but not the seed itself. (Mendeleev 1891, 23)

In this way Mendeleev was able to explain that the only two allotropes of carbon that existed in his era, namely diamond and graphite, both belonged in the same position in the periodic table. Said in other words, the position for carbon in the periodic table does not represent diamond or graphite but carbon in general or carbon in the more abstract sense of a "basic substance."

This issue is of great importance to the scientific episode that Wray is proposing to call a scientific revolution. When isotopes were first discovered they did indeed pose something of an existential threat to the periodic table because it appeared as though the number of elements was very rapidly proliferating and it was by no means clear where, if at all, they should be accommodated into the periodic table.

The person who was responsible for solving the problem was the philosophically astute radio-chemist Friedrich Paneth (Paneth 1962). It was Paneth who provided the philosophical and conceptual underpinning that paved the way for the new official definition of an element that was adopted by the International Union of Pure & Applied Chemistry (IUPAC) in the 1920s. I do not believe that such a change was primarily one concerning the lexicon of chemistry. It was rather a fundamental reinterpretation and rationalization of the isotopes that were being discovered. Moreover, Paneth realized that the periodic table could safely ignore this apparent proliferation since it was focused on elements as basic substances, as characterized by their atomic numbers.

Let me turn to another aspect that I believe Wray fails to address. When Kuhn revised his earlier views on scientific revolutions in order to focus on the lexicons in science, he also turned from the consideration of diachronic changes to those that were synchronic and involving the diversification of subdisciplines. What Wray appears to be doing is adopting Kuhn's later definitions of paradigm and revolution while still considering the development of science across time. For the later Kuhn, paradigms take on a positive role in the form of the development of subdisciplines and specialties, as Wray has eloquently written about in his book on Kuhn (Wray 2011). Kuhn compares the diversification of scientific fields to that of biological speciation. As Kuhn sees it, the form of incommensurability that develops when two disciplines branch off from each other is actually productive. Divergences in lexical taxonomies are supposed to lead to communication difficulties among members of rival paradigms and the later Kuhn believes that such isolation is favorable to the growth of science.

It is by no means obvious how this aspect of the later Kuhn's thinking is supposed to apply to Wray's example since, as I have stated earlier, it was not so much a matter of changing the meaning of "element" as one of favoring a meaning that had been largely neglected since the time of Mendeleev. It was also a matter of focusing on the nucleus of any element rather than the atom as a whole, as I explained earlier.

For the later Kuhn speciation, and the resulting isolation, enables scientific disciplines to specialize and science to grow as a result. Wray makes no such argument for what he is claiming to be a scientific revolution, in the sense of the later Kuhn, regarding the change from atomic weight to atomic number. It would seem that he has concentrated on one aspect of the later Kuhn's view, namely lexical changes, while not also adopting the focus that Kuhn recommended on synchronic changes such as those involving the diversification of scientific subdisciplines, that is to say, the positive slant that Kuhn gives to his new interpretation of incommensurability. In this respect Wray may be supporting a rather outdated Kuhnian view while correctly reporting that the newer conception of elements represents an improvement on the view in which elements were characterized by their atomic weights.

7.7 Kuhn's Non-Overlap Principle

In order to explain his revised view of incommensurability and the new sense of scientific revolutions in terms of changes in lexicon, Kuhn also proposed his no-overlap principle in the course of an address delivered to the Philosophy of Science Association in 1990 (Kuhn 1990).

According to Kuhn there is a revolution if there is no "overlap" between two kind terms such as the term "planet" as it was used by Ptolemaic and Copernican astronomers respectively. For Ptolemy, and his followers, the Sun was regarded as a planet in view of its apparent wandering motion in the sky. On the other hand, the Earth was not regarded as a planet. For Copernicus, and his supporters, it was the other way around since the Sun was not thought of as a planet whereas the Earth was.

Although Kuhn never presented diagrams to depict this state of affairs I will attempt to do so later for the sake of clarity (Figure 7.1).[10]

The intersection of the two sets contains the celestial bodies that both groups of astronomers regarded as planets. In the everyday sense of the word, one could say that there is considerable overlap among the members of these two sets of celestial bodies. The problematic cases clearly consist of the Sun, the Earth, and the Moon. However, I do not see that these three anomalies would present serious difficulties.

On a strong reading of Kuhn, a lack of overlap would seem to indicate a lack of complete overlap since the kind terms involved in the taxonomy can only stand in two relations, namely exclusion or inclusion. Partial overlap between contrasting kinds is forbidden in that something is either a planet or a star but not both.

I now turn to how Wray puts Kuhn's no-overlap principle to use in arguing for his claim that the change in the definition of an element represented a scientific revolution. Wray says that if the defining feature of a chemical element is its atomic weight, then isotopes are strictly speaking impossible or forbidden. According to Wray, the earlier lexicon regarding elements was incompatible with the very possibility of two

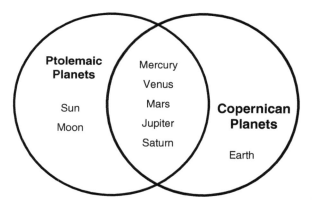

Figure 7.1 Comparison of the extension of the term "planet" in the Ptolemaic theory and the Copernican theory. Both groups of astronomers regarded Mercury, Venus, Mars, Jupiter, and Saturn as planets.

[10] As Wray tells me, there is very little written by Kuhn on the no-overlap principle. Many authors and commentators on Kuhn appear to merely repeat what Kuhn said using the same examples that include cats and dogs as well as gold and silver rings, between which pairs Kuhn also claims that no-overlap exists.

samples of the same element having different atomic weights. The only way to accommodate the possibility of isotopes, continues Wray, is to radically change the lexicon by defining elements through their atomic numbers instead of their weights. As a result, samples of the same element need not necessarily have the same atomic weight.

However, I do not think that the two cases, Kuhn's example of planets and Wray's example of elements, are analogous, because any substance that has an atomic weight also has an atomic number. Consider the classic example of the two major isotopes of chlorine, namely ^{35}Cl and ^{37}Cl. Any atom of ^{35}Cl contains its atomic number within itself, as it were, since the nucleus of this isotope contains seventeen protons. Similarly, every atom of the other isotope ^{37}Cl also contains a nucleus with seventeen protons. An atom cannot be an isotope of any particular element without containing within itself the feature that identifies it as an isotope of that particular element, namely its requisite number of protons.

The simple reason for this state of affairs is that ontologically speaking atomic number is fully contained within the weight of any particular isotope. Stated otherwise, atomic number is associated with just the protons in an atom, whereas atomic weight is due to the number of protons plus the number of neutrons and electrons. Focusing on atomic number, as is the case in the modern definition of an element, implies focusing on a more specific aspect of the same ontological entities, or the atoms themselves, regardless of whether they might be the same or different isotopes of any particular element.

I now return to my Venn diagrams, this time applied to elements, atomic number and atomic weight, and the relationship between them. One way to represent this relationship is shown in Figure 7.2.

So even if Kuhn may be correct in insisting that there is no *complete* overlap in the case of planets as viewed through the two lexicons, Wray may not be entitled to make an analogous argument in the case of elements viewed through the lexicon of atomic number as opposed to atomic weight.

Not only is there overlap in the element case but there is also complete containment of one set within the larger one, ontologically speaking. To cope with the discovery of isotopes scientists were required to sharpen their focus and their definition of element-hood. They were not required to exclude certain instances of previously regarded elements. There was no counterpart to Ptolemy regarding the Sun as a planet that then became reclassified as a star in the subsequent lexicon. This is why I believe that the two situations are not analogous. As a result, and in

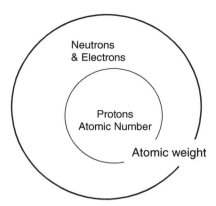

Figure 7.2 Venn diagram of the relationship between atomic number and atomic weight. Atomic number (protons) is fully contained within atomic weight (protons plus neutrons and electrons) for any particular isotope.

the terms of the later Kuhn, the change in lexicon from using atomic weight to using atomic number to define elements does not, in my view, constitute a Kuhnian revolution, neither classic nor in the sense of the later Kuhn.

7.8 Conclusion

Contrary to Wray's ingenious proposal I do not share the view that the discovery of atomic number as the ordering principle for the elements represented a scientific revolution. As I have argued, it was certainly not a revolution in the sense of Kuhn's original account in SSR. Moreover, I do not believe that it was a revolution in the sense of the later writings of Kuhn in which attention shifts to the scientific lexicon, the diversification of scientific fields, and the no-overlap principle.

Finally, I must raise another issue that occurred to me after I had essentially completed this article. In all that I have written here, I have examined Wray's view that the adoption of atomic number by the chemical community represented a scientific revolution. But there are really two issues at play here. First there is the question of whether Wray has correctly identified a historical episode that has the characteristics of what Kuhn might consider to be a scientific revolution. Broadly speaking I believe he has done so. The other question is whether this episode does indeed represent a scientific revolution *tout court*, regardless of Kuhn's views. I am a little worried that I may have allowed my deep-seated disagreement with Kuhn, specifically on revolutions, to be transferred to one of his followers, namely Brad Wray (Scerri, 2016). Needless to say, this is somewhat inevitable since Kuhn is no longer with us, and all I can do is to direct my comments at one of his finest living expositors. Thank you, Brad Wray, for graciously allowing me to contribute to this volume. If my chapter appears like a fish out of water, it may be because I choose to focus on the scientific details. In the final analysis I would hope that Kuhn might approve of this course of action.

PART III
Kuhnian Themes

CHAPTER 8

The Copernican Revolution since Kuhn

Peter Barker

8.1 Introduction

How has our understanding of the Copernican Revolution changed since Kuhn's book about it in 1957? Although, like Imre Lakatos, I believe that history of science without philosophy of science is blind (Lakatos 1971, 91), the modern historical image of Copernicus and his work is so different from that presented in CR that a good deal of ground work needs to be laid before any sort of philosophical discussion can begin. Consequently this chapter is divided into two main parts: I will present a historical narrative first, followed by a philosophical discussion. I will argue that historical information on the relationship between Copernicus's work and Islamicate astronomy, which first came to light at essentially the time Kuhn was writing CR and SSR, very much complicates the depiction of Copernicus's work as revolutionary or discontinuous with previous astronomy. However, the situation in European astronomy and cosmology from the time of Copernicus to the death of Newton does look like what Kuhn described as a crisis state. It is also possible to locate incommensurabilities between heliocentric cosmology and geocentric cosmology, especially beginning with Kepler. And the concept of incommensurability remains an important resource for understanding the history of science.

8.2 Situating Copernicus in the History of Astronomy

Copernicus's book of 1543 appeared as a radical departure in astronomy to Northern Europeans. It not only shifted the effective center of the planetary system to the Sun, but the mathematical presentation seemed to be completely original, especially in Copernicus's avoidance of the Ptolemaic mathematical device known as the equant. Although very few people were convinced that Copernicus's theory was physically true, in the period from its publication to the careers of Kepler and Galileo, many people admired the mathematical techniques he introduced. Copernicus himself was clear that shifting the center of the planetary system to the Sun was not a new idea, a point also reflected in Kuhn's work (CR 42; SSR-2 75). But it has since become clear that his mathematical methods were also not really innovations. Copernicus borrowed them from the much more sophisticated mathematical astronomy available in the Islamicate world (the parts of the world governed by Islamic rulers) before and during his life time, but without acknowledging his sources. Ironically the first important papers on the connections between Copernicus and Islamicate astronomy began to appear at exactly the same time that Kuhn was completing CR (1957) and writing SSR (to mention only some of the most important: Roberts 1957; Kennedy and Roberts 1959; Kennedy 1966; Hartner 1969; Kennedy 1971; Hartner 1973). To understand this background to Copernicus, let me summarize the current consensus on the history of astronomy in the Islamicate world.

By the second century after the life of the Prophet, bureaucrats and court officials in the Abbasid Empire (750–1258) highly valued mathematical astronomy. In an empire that extended from central Asia to Western North Africa, one urgent question was determining the correct direction for prayer and the alignment of religious buildings like mosques. The direction to Mecca, or *qibla*, could be determined by various rules of thumb, but as Muslims traveled across a territory stretching from Northern India to Spain, a precise and definitive means of establishing the direction became increasingly urgent. Mathematical astronomy also formed the basis to timekeeping, including the timing of daily prayers and the dates of annual festivals. The growing empire also needed an accurate calendar for collecting taxes. Last, and by no means least, all levels of society, but especially Islamicate rulers, were addicted to astrological advice (Saliba 1992). All these needs were served by the production of astronomical handbooks, or *zij*s. Early *zij*s drew on sources from Persia and India. They also drew on the Greek tradition including the work of Claudius Ptolemy (c. 160) preserved in Constantinople and elsewhere on the border of the Abbasid Empire (King, Samsó and Goldstein 2001). Under the caliph al-Ma'mun (r. 813–833), Islamicate astronomy became almost exclusively Ptolemaic (Sayili 1960, 79–80). In Kuhnian terms we might say, without much distortion, that a paradigm emerged, with a central text, and a normal science tradition devoted to making *zij*s for different regions, as well as making and improving the instruments needed to make the observations used in making a *zij*.

The central text of the paradigm was Ptolemy's *Mathematike Syntaxis* (composed c. 160), known in Arabic as *al-Majisti* and in Europe, later, as the *Almagest* (Toomer 1998). It gave a complete set of mathematical techniques for calculating planetary motions and even described how to construct some of the main instruments needed for making observations to determine key constants. The first Arabic translations became available around 800; however, Ḥajjaj ibn Yusuf ibn Maṭar (786–830) translated Euclid's *Elements* for al-Ma'mun and also produced a new translation of the *Syntaxis* (Brentjes 2007, 460). The caliph sponsored observatories at Shammasiya and Qasiyun near Baghdad. At these observatories new versions of traditional instruments were used to make new and more accurate observations, for the purpose of making a new *zij* (Sayili 1960, 50–87).

Typically a *zij*, or astronomical handbook, contained an introduction to the mathematical techniques needed to use it and sets of tables from which the positions of the Sun, the Moon and the other five known planets could be worked out without needing to go back to first principles calculations using eccentrics and epicycles. Al-Ma'mun's tables, known as the *Tested Tables*, set a new standard for accuracy. All this was achieved according to methods laid out by Ptolemy in the *Syntaxis*. Normal science under this paradigm consisted of making observations using instruments described by Ptolemy, as well as those inherited or newly devised, such as astrolabes, and using Ptolemy's methods to construct the tables recorded in *zij*s.

As specified by Kuhn (SSR-2, chapter 3), normal science at the time of al-Ma'mun refined the values of key constants such as the length of the solar year, and the obliquity of the ecliptic and the eccentricity of the Sun. This series of observations led to the recognition that the apogee of the Sun moved over time, a result unknown to Ptolemy (Sayili 1960, 77). The novelty did not undermine the paradigm because it changed a known constant to a variable that could be measured by established techniques. Other efforts attempted to improve the layout of tables, and simplify their use, as well as making new instruments and improving old ones (Sayili 1960, 72–73).

By the tenth century a new genre of astronomy book began to emerge, called *hay'a*, which gave a systematic description of the structure of the cosmos (King, Samsó and Goldstein 2001, 70–73; Ragep 2008). The first part of these books usually described a central Earth composed of the spheres of the four elements recognized by Aristotle. The structure of the heavens was presented in the order used by Ptolemy (Moon, Mercury, Venus, Sun, Mars, Jupiter, Saturn, fixed stars). However, from the beginning of the genre the motions of the planets were described by means of sets of three dimensional orbs that carried the planets as they rotated. As in Ptolemy's mathematical models, these orbs contained physical elements identified as eccentrics and epicycles, which varied for the Sun, Moon and Mercury. Venus and the remaining planets represented different versions of a fourth pattern (Ragep 2008). The basic ideas for these physical models had been provided by Ptolemy himself in the *Planetary Hypotheses*, written immediately after the *Syntaxis* (Goldstein 1967). This book remained unknown in the Latin West, where the orb models were simply attributed to Islamicate authors, who had also provided most of the information contained in the two basic Western astronomy texts, the *Sphere* of Sacrobosco (Thorndike, 1949; Pedersen 1985; Crowther, et al., 2013) and the anonymous *Theorica planetarum* (Pedersen 1978, 1981). It is important to underline the relative poverty of Western astronomy before the time of Copernicus. Although individuals like Roger Bacon (d. c. 1292) in England and Bernard of Verdun (fl. c. 1300) in France understood Islamicate orb models (Grant 1996, 279–281; Bernard of Verdun/Hartmann 1961), neither the *Sphere* nor the *Theorica* introduced them. As a result, Western astronomers simply did not participate in the advances made by Islamicate astronomy after the eleventh century.

By the mid-eleventh century astronomy in the Islamicate world had progressed considerably beyond its Greek origins. As indicated earlier, many important constants had been measured with greater precision, and the motion of the Sun's apogee had been discovered and added to the models for the motion of the Sun. However, the increasing expertise of Islamicate practitioners also caused them to recognize problems in Ptolemy's account – anomalies for the Ptolemaic normal science tradition, if you will. Chief among these were problems caused by mathematical techniques used by Ptolemy that could not be translated into motions of orbs. For example, in his Moon model Ptolemy separated the geometrical center of the Moon's eccentric from its center of rotation, and also required the apogee of the Moon's epicycle to rock back and forth, using a device he labelled *prosneusis* (Pedersen, 1974: 192–195; Saliba, 1994: 137–138). In the models for the planets, Ptolemy again introduced a separation between the geometrical center of the eccentric and its center of rotation. The latter point was named the equant and could not plausibly be fitted into sets of orbs that rotated about their diameters (which, of course, ran through their geometrical centers).

Ibn al-Haytham (Latin: Alhazen, d. Cairo, 1040) wrote a book pointing out and listing all the difficulties in reconciling the *Syntaxis* with physical

The Copernican Revolution since Kuhn

orbs (Ibn al-Haytham/Sabra 1971; Voss 1985). These became the central theoretical problems of Islamicate astronomy for the next two centuries, but, as far as we know, subsequent work was not transmitted to the West (Voss, 1985, 4), and a similar research program never developed there. *Pace* Kuhn, it should be pointed out that these problems are instances of normal science puzzle solving. Ptolemy introduced the lunar equant and prosneusis because a simple concentric-epicycle model could not predict the position of the Moon at quadrature and octants. But the objection that equants and motions governed by prosneusis cannot be recovered as motions of rotating orbs is not the kind of anomaly recognized by Kuhn in SSR, which is a discrepancy between theory and observation (SSR-2 52–53; Andersen, Barker and Chen 2006, 69ff.). Rather these are internal or conceptual problems that for the most part have no direct bearing on the theory's empirical success (Andersen, Barker and Chen 2006, 139–146). A possible exception was another problem of the Moon model. The combination of circles used by Ptolemy required the Moon to come dramatically closer to the Earth at quadratures, which should have resulted in a large change in the size of its visible diameter twice over a month.

The first important solutions (that we know of) to these problems were provided by a single group of astronomers gathered by the Mongol conqueror Hulegu Khan, who destroyed the remains of the Abbasid caliphate in 1258. At an observatory that he commissioned near his capital Maragha in northern Persia he assembled the most famous astronomers in his now wide realm, including Muyyad al-Din al-Urdi (d. 1266), from Damascus, and Nasir al-Din al-Tusi (d. 1274), who he recruited from the Ismaili stronghold of Alamut. With Tusi in charge, this group built the observatory, made new observations and wrote a new *zij*, named for its first patron and his successors as the *Ilkahnid zij*. The same group also provided not one but two major solutions to the equant problem.

Tusi had already invented a mathematical device – now known as a Tusi couple – that used two circles, or spheres, to produce a straight line motion (https://people.sc.fsu.edu/~dduke/ntusi.html). Used as a replacement for Ptolemy's epicycle, this system, in effect a double epicycle, permitted the construction of planetary models in which the orbs all rotated about their geometrical centers. Urdi demonstrated a second method, which redefined the eccentricity of Ptolemy's model and introduced a small auxiliary epiclet. This system also allowed the construction of models with orbs that rotated uniformly about their geometrical centers (Figure 8.1).

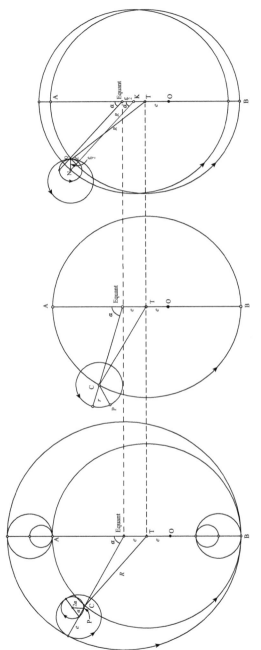

Figure 8.1 Comparison of Ptolemy's, Tusi's, and Urdi's models for an outer planet.

Neither Tusi nor Urdi provided solutions to all the outstanding problems, but they provided an arsenal of methods that encouraged later astronomers to develop new models based on their techniques. Tusi's model for the Moon removed the lunar equivalent of the equant and the prosneusis device in favor of well-behaved circles or spheres that rotated about their geometrical centers. But the new model did not solve the distance variation problem inherited from Ptolemy – it still required the Moon to double in size twice a month (King, Samsó and Goldstein 2001, 73–75).

From the death of Tusi in 1274, advanced work in Islamicate astronomy shifted to approaches using the methods developed at Maragha by Tusi and Urdi, although Ptolemy's original models continued to be used and to be covered in introductory works (King, Samsó and Goldstein 2001, 53). This change was so dramatic that in 1987 George Saliba, perhaps the most eminent authority on Islamicate astronomy by the end of the twentieth century, described the adoption of Tusi and Urdi's methods in a paper entitled "The Role of Maragha in the Development of Islamic Astronomy: A Scientific Revolution Before the Renaissance" (Saliba, 1987). Later, I will consider whether or not these events deserve to be counted as a revolution in Kuhn's sense. Some of the most important astronomers who worked in this new tradition were Ibn al-Shatir (d. 1375), an important instrument builder and theoretician from Damascus, and Ali Qushji (d. 1474) who helped to bring the Maragha techniques from Samarkand to Istanbul.

Ibn al-Shatir was the timekeeper at the Ummayad mosque in Damascus for most of his life. His official duties were to use astronomical means to determine the times of religious observances, starting with daily prayers and extending to major festivals and the annual calendar (King 1975, 360–361). Ibn al-Shatir made at least two new *zijs*. The first used Ptolemy's techniques but the second adopted and extended Maragha methods (King 1975, 358–360). It had been known since antiquity that any eccentric circle (or orb) may, in principle, be replaced by a concentric circle (i.e., one centered on the Earth) carrying an epicycle. Ibn al-Shatir used this theorem plus Maragha techniques to construct an astronomy completely without eccentrics. Most of his models feature double epicycles. His model for Mercury, always a problem because of its large eccentricity, added a Tusi couple to periodically lengthen the radius of the final epicycle (see image in King 1975, 359). He also improved on Tusi's Moon model and for the first time (as far as we know) solved the distance/size variation problem. By combining two epicycles, and placing the Moon on the second one, the outward displacement created by the first could be almost eliminated by the second, creating a model that not only gave correct longitudes but also predicted no great size variation during the Moon's monthly passage around the Earth (see image King 1975). Although Ibn al-Shatir lacked the extensive support and colleagues available at Maragha or Samarkand his non-Ptolemaic zij was widely used and his planetary models also circulated in manuscript (King 1975, 2007).

Between 1416 and 1420, Ulugh Beg (d. 1449), the ruler of Samarkand, and a prince practitioner in his own right (Moran 1981), founded a *madrasa* or college devoted to the mathematical sciences and built an observatory on nearly the same scale as Maragha (Van Dalen 2007). He also collected experts to work there, though, unlike Hulegu, he used persuasion not compulsion. This group continued the normal science tradition by making new observations and constructing a new *zij*, named after their patron (King, Samsó and Goldstein 2001, 54).

Ali Qushji, who was born in Samarkand, studied at Ulugh Beg's *madrasa* and later became director of the observatory. He used Maragha methods to construct new models for planetary motion, including the very problematic case of Mercury (Fazlıoğlu 2007). He also explicitly examined the possible motion of the Earth, which he did not endorse, but allowed as a logical possibility (Ragep 2001). When he left Maragha after the assassination of Ulugh Beg, he was already sufficiently famous to obtain positions at a succession of major courts, ending at the court of Ottoman sultan Mehmed II, who appointed him to a major position at the Hagia Sophia mosque. Qushji died in Istanbul in 1474, two years after the birth of Nicholas Copernicus in Torun, Poland (Barker and Heidarzadeh 2016). Several generations of Qushji's children and students continued his astronomical work in the Ottoman Empire.

It is a myth that science in the Islamicate world declined and vanished just as science arose in Europe. The events offered to date the collapse of Islamicate science vary, but all are refuted by the history of astronomy. The first is usually the career of al-Ghazali (d. 1111), who is supposed to have persuaded Muslims to abandon science on religious grounds. He did not. Rather he rejected Aristotelian metaphysics and especially the concept of cause. His critique led to the founding of new occasionalist and atomist viewpoints. In fact he was a strong supporter of mathematical sciences including astronomy (Griffel 2009; Griffel forthcoming). The second event offered is the extinction of the Eastern caliphate in 1258, but, as we have already seen, the conqueror Hulegu Khan established the Maragha observatory, which led to a whole new phase of astronomy in the Islamicate world. The final date offered for the alleged collapse of Islamicate science is the death of Ulugh Beg in 1449. But astronomers trained at Samarkand and their students did not just migrate West to Istanbul and the long-lived Ottoman Empire (c. 1300–1923), they also went to the regions of Persia that became the Safavid Empire (1501–1722). Here, based on the work of Shams al-Din Khafri (d. 1550) and others, Baha al-Din al-Amili (d. 1621) founded a new school of astronomers continuing the Maragha methods. His students and successors extended the normal science tradition for at least two centuries and carried it to the Mughal Empire (1526–1739) (Brentjes 2010). The main business of normal science in the Islamicate astronomical tradition continued to be building new instruments and making new *zijs*. In fact (in our current state of knowledge) we might date the last important example of normal science in this tradition to the career of Maharajah Jai Singh of Amber (1688–1743) who between about 1720 and 1740 constructed no less than five monumental observatories in Delhi, Mathura, Ujjain, Benares and his own newly built capital of Jaipur. He also constructed a new *zij*, initially from traditional sources but revised on the basis of methods imported from Europe (Pingree 1999) and sponsored the translation of Tusi's Moon model into Sanskrit, for the benefit of Hindu astronomers (Kusuba and Pingree 2002). Islamicate astronomy was alive and well up to the time that the American colonies became the United States. However, the goal of this section is not to convince you that the rise of Western Science has been overstated and misdated but to contextualize the work of Nicholas Copernicus, treated by Kuhn as epoch making.

8.3 What Copernicus Really Said

Beginning at the same time that Kuhn was writing first CR and then SSR a series of papers showed that fundamental ideas in Nicholas Copernicus's work had all appeared earlier in Islamicate astronomy. It was pointed out, first, that Copernicus's Moon model – with its characteristic double epicycle – was astonishingly like Ibn al-Shatir's, as was his model for the planet Mercury. It was also recognized that Copernicus made use of Tusi's device for turning two circular motions into straight line motion. Perhaps the most curious – and much debated – connection appeared in 1973 when Willy Hartner compared the lettering on Copernicus's diagram of the Tusi couple with the diagram in Tusi's most important book and showed that the majority of the letters in Copernicus were translations of the letters in Tusi. This did not mean that Copernicus had access to Tusi – the same diagram appears with Tusi's lettering in many later commentators. However, the suggestion that Copernicus arrived at the same lettering as Tusi by accident is easily dispelled by examining the next few European

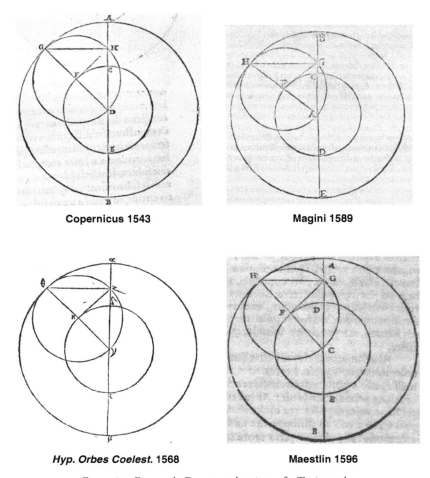

Figure 8.2 Four early European drawings of a Tusi couple.

diagrams of the Tusi couple. All of them derive from Copernicus's figure, but none of them have the same lettering as Copernicus's version, or each other (Figure 8.2).

Tusi had introduced the couple to construct planetary models that avoided the equant mechanism. Copernicus became the first European astronomer to voice the same concern (as far as we know).

Copernicus presented his planetary theories twice, first in a short summary with little mathematics, now known as the *Commentariolus*, and almost certainly completed before 1514, and second, in his main book the *De revolutionibus*, which appeared in 1543 (Swerdlow 1973; Copernicus/Duncan 1976). Both books employ Tusi couples to produce the precession of the equinoxes, to produce the latitude variation in the planets and as the outermost element in the model for Mercury, which copies Ibn al-Shatir. He also copies Ibn al-Shatir's Moon model. In both books Copernicus begins by stating objections to the equant. In the first book he uses Ibn al-Shatir's method for avoiding it in his models for the planets; in the second, he switches to Urdi's alternative method. Thus, as a minimum, Copernicus's astronomical work includes material that duplicates work by Tusi, Urdi and Ibn al-Shatir, not so much as a continuation but as a partial assimilation. None of this is acknowledged by Copernicus. Regardless of whether you accept that Copernicus's diagram of the Tusi couple is a direct copy of the original, the correspondences between Copernicus's work and the Islamicate tradition in astronomy were so great that by the 1980s almost all experts had concluded that "[t]he question therefore is not whether, but when, where, and in what form [Copernicus] learned of Marāgha theory" (Swerdlow and Neugebauer 1984, 1:47). The ideas that Copernicus could have discovered all this by himself – the last gasp of Great Man historiography – was dramatically implausible.

In addition Ragep has argued that what is often seen as Copernicus's key innovation in method – the liberation of astronomy from Aristotle's physics – was also prefigured in the Islamicate world and especially in the work of Qushji (Ragep 2001). I believe Ragep is correct here, although I have suggested elsewhere that the Islamicate separation of astronomy and traditional physics appears in Europe a little earlier than Copernicus and forms a background to his work (Barker 2011b; 2013b). Qushji also forms part of a long tradition of Islamicate discussions of the rotation of the Earth, and Copernicus treats comets identically to Qushji (Ragep 2001). We should recall here that Qushji had been at the court of the Ottoman sultan in Istanbul at the time of his death in 1474 and that Copernicus began his education in Italy only twenty-two years later.

In 2014 Robert Morrison offered an attractive intermediary: Moses Galeano (d. after 1536), a Jewish scholar who had worked at the sultan's court in Istanbul, and who was visiting Venice at the same time that Copernicus was studying next door at Padua. All the material that turns up in Copernicus was readily available in Istanbul, and indeed, Moses Galeano himself later wrote astronomy books using large parts of it (Morrison 2014). Perhaps Copernicus heard all these theories as part of a private lecture series or a series of briefings by Galeano. That might also explain why he does not name any of the originators of the mathematical methods he uses. He may never have been told, or he may have assumed that they were all the work of Galeano or his contemporaries. According to sixteenth-century scholarly practice one did not name living authors.

During Copernicus's career the situation in European astronomy was dramatically different from the situation in the Islamicate world. In the East technical astronomy followed Ptolemy or his successors in the Maragha school. It was not that Maragha ideas never reached the West – they seem to have arrived repeatedly and by different routes – rather they never found a purchase (Barker and Heidarzadeh 2016). Even Copernicus did not succeed in getting them accepted into the astronomical mainstream; Maragha methods remained marginal until heliocentrism was reframed by Kepler and Galileo, who abandoned Copernicus's mathematical models.

At the beginning of Copernicus's education in Italy, mathematical astronomy in Europe reached a new level or proficiency. Two landmark books are the 1474 edition of the *New Theorica of the Planets* by Georg Peurbach (1423–1461), published in Nuremberg by his student Regiomontanus (1436–1476), and the posthumous appearance of Regiomontanus's *Epitome of the Almagest* in 1496. But technical astronomy – astronomy capable of making predictions – became newly embroiled in a longstanding dispute with natural philosophers who followed the Andalusian philosopher Averroes (Ibn Rushd, 1126–1198). The latter had flatly denied the existence of eccentrics and epicycles as anything other than mathematical fictions, while the whole point of the *New Theorica* was to found the planetary models on physically real orbs, as had always been the case in the Eastern Islamicate world. Averroes's critique never took hold in the East, but it became unavoidable in the West, where Averroes became the authoritative interpreter of Aristotle and one of the foundations of the university curriculum (Hasse 2016). Copernicus and his peers were therefore obliged to defend their version of astronomy, with its orbs, eccentrics and epicycles, from continuous attacks by natural philosophers. In a famous quotation Copernicus makes his position quite clear:

> For while some employ only orbs centered on the earth, others use eccentrics and epicycles, and yet they do not quite reach their goal. Although some people who adopted earth-centered orbs showed that in this way some non-uniform motions could be explained, by this means they were unable to obtain any results that were both certain and agreed with observation. On the other hand those who use eccentrics seem to have largely solved the problem of the apparent motions by calculation. But at the same time they introduce concepts that contradict the first principle of uniform motion.

It was the Averroists who "were unable to obtain any results that were both certain and agreed with observation"; that is, they could not make numerical predictions. It was the followers of Ptolemy and Peurbach's *New Theorica* who "contradict the first principle of uniform motion" by using the equant. Copernicus sides with the latter and solves their problems by drawing on the tools developed in Maragha. The dispute with the Averroists continued after Copernicus's death and was finally resolved by Tycho and Galileo, as we will see later. His avoidance of the equant was attractive to some of his contemporaries, but his cosmic scheme was not.

The number of people who accepted Copernicus's heliocentric cosmology before the seventeenth century was around a dozen (for different scores see Westman 1980, 136, n.6, and Tredwell and Barker 2008). Copernicus's own reputation after his death was considerable – his ideas penetrated the arts and politics as well as universities – but did not ensure their acceptance (Yates 1947, 95–101; Strong 1973, 105 fig. 84; Hutcheson 1987; Kaufmann 1993). This was a world where every university student had to take astronomy, so the potential audience was in the tens of thousands, at least. Although Copernicus's mathematical methods were adopted, especially among Lutherans, his cosmic scheme was rejected for physical and theological reasons (Westman 1975; Barker 2005). Kuhn was mistaken in thinking that there was "bitter ... Protestant opposition" to Copernicus's work (CR 193). In fact Lutherans were essential in producing *De revolutionibus* and in spreading its ideas (Barker and Goldstein 2003; Barker 2004b). And in the period from Copernicus to Newton, religion was generally a positive force in scientific change (Barker 2000; Barker and Goldstein 2001; Iliffe 2017).

The Lutherans and others who spread Copernicus's ideas supported Ptolemy, eccentrics and epicycles, and opposed the followers of Averroes (Barker 2009). They were followers of Ptolemy and Peurbach who constructed the heavens with eccentric orbs carrying spherical epicycles. The Averroists continued to insist that the celestial orbs must all be concentric to the Earth. The conflict between them was not resolved by one side winning and the other losing. Rather, following Tycho Brahe, astronomers had the option of abandoning solid spheres and orbs. This change was neither rapid nor complete, but it was the ultimate solution to the long-standing dispute.

Tycho abandoned solid orbs after using his unusually large (and hence accurate) instruments to track the movements of comets and to determine that Mars approached closer to the Earth than the Sun. Unable to accept Copernicanism for the usual Lutheran reasons – its incompatibility with accepted physics and theology – in 1588 he introduced a hybrid system in which the Sun, Moon and fixed stars went around a central Earth, but the remaining planets went around the Sun. This was not a rearguard action. His system combined the best features of Ptolemaic and Copernican astronomy while avoiding the liabilities of both. As late as the 1650s, honest appraisals of the three alternatives found Tycho's system superior on scientific grounds, and it provided an ongoing refuge for intellectually respectable Catholics after the condemnation of Copernicanism in 1616 (Graney 2015; Lernert 1995; Lernert 2008 2:109–116).

One of the main conclusions of recent historical work on Copernicus's career and its aftermath recognizes the discontinuity between Copernicus's goals and agenda, and those of his successors, whether or not they called themselves Copernicans (Westman 2011; Omodeo 2014). This is most striking in the case of Kepler (1571–1630). In his 1611 book *New Astronomy Based on Causes*, Kepler reintroduced the equant, the main announced motivation for Copernicus's new mathematical models (Donahue 2015, chapter 4). He then abandoned the canonical circles of astronomy in favor of ellipses with the Sun at one focus and the empty focus as the equant's ghost. In many other respects Kepler repudiated Copernicus's work – for example his principled rejection of epicycles. He also repudiated the celestial orbs that Copernicus had retained and offered an explanation (the 1596 *Mysterium cosmographicum* construction of Pythagorean solids) for the embarrassing and inexplicable spacing of Copernicus's planets (Barker 2002). However, Kepler was not a prototype mechanical philosopher with mystical lapses. Rather he is a clear example of another historical theme absent in Kuhn's treatment of the Copernican Revolution – the positive role of religion (Barker 2001; Barker and Goldstein 2001). The attitude of historical actors up to and beyond the time of Newton was not that religion had no place in science, but rather that religion and science must naturally support each other. Kepler, for example, believed that he had uncovered God's physical plan for the world, and his first book was an extended argument that God was a Copernican (Barker 2001; Rothman 2017). At the same time his cosmos embraced major spiritual features. Planets were living creatures with souls, and capable of moving themselves radially nearer or further from the Sun, although their orbital motion (Kepler introduced the term "orbit") was caused by a force emitted by the Sun (Boner 2013; Kepler/Donahue 2015, esp. chapters 33–35). The celestial spheres were reduced to geometrical boundaries in an all-pervasive celestial fluid and became largely irrelevant

The Copernican Revolution since Kuhn 159

after the *New Astronomy* introduced new algorithms for calculation with a new speculative physical basis.

Finally, Galileo provided telescopic evidence that Jupiter had moons; Venus showed phases; the Sun rotated and had ephemeral spots; and Earth's moon had mountain, valleys and craters (Galilei and Scheiner/Reeves and Van Helden 2010; Galilei/Van Helden 2016). Despite his lifelong advocacy, none of this proved that Copernicus was right, although it did show that certain versions of Aristotle and Ptolemy were wrong. Whatever the true center of celestial motions, the Earth or the Sun, the moons of Jupiter did not revolve around either of them, so the Averroist principle that all celestial motions must be centered on the Earth was clearly false. As for Venus (and by parallel reasoning Mercury) its motions had to be centered on the Sun. But both these pieces of evidence could be explained equally well in Tycho's cosmic scheme. The strongest conclusion that could emerge was that Aristotle and Ptolemy were wrong, not that Copernicus was right. Galileo's omission of Tycho's alternative from his 1632 book *Dialogue on the Two Chief Systems of the World, Ptolemaic and Copernican* (Galilei/Drake 2001) was a tacit admission that he had no cogent response. In 1616 Copernicanism was condemned for theological *and* scientific reasons. Research in astronomy and physics as late as the decades after Galileo's death – for example, careful Jesuit experiments on falling bodies – failed to support Galileo or vindicate Copernicus (Graney 2015; Dinis 2017).

Copernicanism was ultimately vindicated by Newton, but neither cosmology nor the ontology of physics had reached anything like its modern form. As a twentieth-century physicist, Kuhn was intimately familiar with the great changes in science at the beginning of the century. Both quantum theory and relativity theory produced enormous changes in the ontology of physics over a period of mere decades. And these changes were largely the concern of experts in physics and appeared independent of other scientific fields or causes outside physics itself. Furthermore, these changes used a largely stable base of communication technologies and scientific institutions, from university education and learned societies, to research journals. In retrospect, Kuhn's attempt to project this kind of scientific change back in time seems problematic. Religion interpenetrated the work of sixteenth- and seventeenth-century scientists. Their work drew on research from outside their own culture – and this is especially true of Copernicus – and outside their fields, as defined in modern terms. Their period saw the founding of the communication technologies (printing) and research institutions (societies and journals) that had become standard by the beginning of the twentieth century. Was the work of Copernicus the locus of a revolution or just one cause of these wider changes?

8.4 Was Copernicus's Work Revolutionary?

To appraise the status of Copernicus's work let us first examine George Saliba's provocative claim that the work of Tusi and Urdi in thirteenth-century Persia was a scientific revolution (Saliba 1987). Kuhn's cyclical model of the history of science imposes a structure on such events (Hoyningen-Huene 1993). Revolutions are to be preceded by crisis states in which alternatives to the paradigm proliferate. These in turn are to be preceded by anomalies that resist the usual efforts of normal science. So was there a normal science tradition in Islamicate astronomy before Tusi and Urdi, and if so how did they change it?

As I have already argued, from the time of al-Ma'mun the main business of Islamicate astronomy had been preparing *zij*s or astronomical handbooks containing tables for calculating the movements of celestial objects. In service of this goal new observations were frequently made, often from specially built observatories. The mathematical methods used to convert these observations into tables in *zij*s were those presented by Ptolemy in the *Almagest*. These were quite limited. They consisted of eccentric circles (and by implication concentric circles as an alternative), epicycles carried by large circles, what is now called a "crank" – consisting of a small circle carrying a large circle, the prosneusis method used in the case of the Moon, and finally the notorious equant point, which separated the center of rotation from the geometrical center of a circle or orb. The crank was used only in the models for the Moon and Mercury. Ptolemy had opened the question of the physical basis of astronomy in the *Planetary Hypotheses* but had produced no satisfactory solutions to the problems posed by the equant or crank. Islamicate astronomers pursued this goal, most vigorously after Ibn al-Haytham clearly listed the parts of Ptolemy's models that could not be replaced by rotating orbs. But, as already pointed out, these are not standard Kuhnian anomalies.

The intervening period, up to the founding of the Maragha observatory in 1259, may reasonably be described as a crisis state; there was a long-standing unresolved problem with Ptolemy's mathematical models that despite their success as predictive tools, they could not be made consistent with accepted physics, especially with systems of orbs that rotated uniformly only about their proper centers. Tusi and Urdi solved these problems by reducing the number of mathematical techniques used, not by introducing new ones. They eliminated the equant, for example, either by introducing a novel arrangement of double epicycles (Tusi) or by redefining the eccentricity of the large eccentric circle and supplementing it with a very small epicycle, which in turn carried the main epicycle (Urdi) (see again, Figure 8.1). Neither method introduced new fundamental entities – rather astronomers were now able to use the properly rotating orbs they had always wanted to use as the ontology of the heavens. And all this was referred to a central stationary Earth.

Hence, although the models developed at Maragha and after Maragha were not found in Ptolemy's original work, they continued to use a subset of his basic tools and vindicated the ontology of celestial orbs he had introduced in the *Planetary Hypotheses*. Tusi and Urdi introduced no new fundamental concepts; they used existing concepts in new ways – almost a definition of normal science. Although the equant disappeared as an overt feature of their models, both models did the same mathematical work, and an equivalent point could easily be located within them. The new methods also offered easier calculation techniques. Other aspects of normal science remained unchanged. Building observatories, making observations and constructing *zijs* remained astronomers' main activities, but now many of them used or modified the models from Maragha, although some continued to use Ptolemy. Thus the adoption of the new models was also not universal, as required for the contents of a new paradigm. For all these reasons, then the work of Tusi, Urdi and their numerous successors, including Ibn al-Shatir and Qushji, counts as a continuation of Ptolemaic normal science. Tusi and Urdi did not create a paradigm shift.

This conclusion immediately affects our appraisal of Copernicus. Copernicus uses mathematical resources provided by Ptolemy (eccentrics, concentrics and epicycles), Tusi (the "Tusi couple" used for precession of the equinoxes, planetary motions in latitude, and the Mercury model in both the *Commentariolus* and *De revolutionibus*) and Urdi (especially the models for the outer planets in *De revolutionibus*). We have just argued that Tusi and Urdi continued normal science under Ptolemy's paradigm. Ibn al-Shatir, who provided the models for the Moon and Mercury used by Copernicus in both his books, and provided the main planetary models in the *Commentariolus*, is a successor to Tusi and Urdi and continued their normal science tradition using their tools. The main difference, which also appears in Copernicus, is the replacement of some eccentric circles (or orbs) by concentric circles (or orbs) carrying epicycles. Copernicus may be seen as another Ptolemaic normal scientist, but for one difference. He refers the center of celestial motions not to the center of the Earth but to the Sun, in the non-mathematical parts of *De revolutionibus*, and to a nearby point, the mean sun, in his mathematical models. Does this make his views revolutionary?

According to SSR, a new paradigm should be logically incompatible with its predecessor, because it successfully explains an anomaly which contradicted the old paradigm and led it to crisis (SSR-2 chapter 9, esp. 97). In fact the Ptolemaic methods and models do not legislate a central, nonmoving Earth. And the motion of the Earth was discussed within the Ptolemaic normal science tradition. Remarks by Tusi on the motion of comets and their possible role as evidence in discussing the Earth's motion attracted many later detractors and supporters, including his student Qutb al-Din Shirazi (d. 1311), and al-Sayyid al-Sharif al-Jurjani (d. 1413) who wrote commentaries on both Tusi and Shirazi and lived in Samarkand from 1387 to 1405. Finally, Ali Qushji, who brought advanced astronomical theories from Samarkand to the newly founded Islamic schools in Istanbul after its conquest explicitly discussed the rotation of the Earth and allowed it as admissible in astronomy (Ragep 2001). There are parallel examples in Europe, notably Nicole Oresme (d. 1382) (CR 115) and Celio Calcagnini (d. 1541) an Italian humanist and contemporary of Copernicus, who also considered the possible rotation of the Earth two decades before the publication of *De revolutionibus* (Barker 2005, 35–36; Omodeo 2014, 209–213). However, none of these figures discussed the annual motion of the Earth. The annual motion seemed to be ruled out by empirical evidence, the absence of stellar parallax, and, perhaps more importantly by the impossibility of accounting for it using accepted physical theories. Copernicus proposed a new astronomical system. Despite a few suggestive but problematic comments about rotation as a natural motion of spheres or orbs, he offered no physics competent to explain his astronomy.

Copernicus adopted the principle that the natural motion of an orb is to rotate (*De revolutionibus* Bk 1, esp. chapter 4). In Aristotle's system this applies to the matter of the heavens – the natural motion of the substance of the heavens was motion in a circle, so the celestial orbs rotated. But to make the Earth rotate required a new principle – that all material orbs rotate. Copernicus suggested that this principle might explain both the rotation of the Earth and the motions of the celestial orbs. But the principle is massively violated in his system. The largest orb of all, the orb of the fixed stars, is stationery for Copernicus. And at (or near) the center of his system the Sun is also an orb, and in 1543 there was no evidence that it rotated. A further problem is that the orbs required to carry Copernicus's planets would not be in contact but would have large and inexplicable gaps between them (Barker 2000, 85–88; Barker and Goldstein 2001, 94). But the most serious problem is the lack of anything in the way of a universal physics, like that of Aristotle, to support Copernicus's moving orbs.

Starting from Ptolemy's cosmos, the Copernican cosmos can be created by swapping the positions of the Sun and the Earth-Moon system. This spatial rearrangement is not ontologically novel in the sense that it introduces new referents, but it is revolutionary in one of Kuhn's most important senses: it requires the redistribution of known objects across existing categories in previously forbidden ways, with concomitant changes in those categories or concepts. Kuhn's clearest example of this phenomenon is the Chemical Revolution, roughly that episode in which affinity theory and the phlogiston theory of combustion were replaced by Lavoisier's new ontology of elements and compounds, together with his notation for expressing their constituents (SSR-2 59–60, 69–72). Along the way, metals get redistributed from compounds (of phlogiston and different calxes) to elements (which are fundamental substances), and phlogiston vanishes. Oxygen appears as a new substance, but it merely fills out an existing category (elements); it does not introduce a fundamental ontological novelty. Similarly, Copernicus's proposed heliocentric cosmos requires recategorizing the Earth as a planet, recategorizing the Sun as a star and a center of motion, and recategorizing the Moon in a seemingly new category of satellite. We might argue that the category of satellite is not entirely new, because in late antiquity Martianus Capella made Mercury and Venus, but not the other planets, satellites of the Sun. And Tycho Brahe was quick to extend the status of satellites of the Sun to all the planets, leaving only the Sun and Moon centered on the Earth. These changes are diagnostic of a scientific revolution for Kuhn, but in the case of Copernicus they preserve all previous celestial objects; none are added or lost during the recategorization. The major ontological loss from the old cosmos to the new one is the elimination of the celestial orbs, but this does not occur until well after Copernicus and is not anticipated by him. Nor is it discussed by Kuhn.

For all these reasons, then, it is too simple to assign the location of the revolution to Copernicus's work. At best it is a promissory note for a new cosmos, and proposing it raises more problems in cosmology, physics and theology than it solves. Both Kepler and Galileo became convinced Copernicans before they developed the arguments for Copernicanism that converted later followers. They may well have been influenced in part by the symmetry or beauty argument that Kuhn appeals to (CR

171–176), as well as the prospects of developing the system more fully (Lakatos 1978, 49–52). Similarly, the reasons Copernicus became a Copernican remain unsettled; however, one major argument suggests Copernicus held neo-Platonic values of systematicity and unity that could be upheld in a heliocentric universe but not a geocentric one. Kuhn suggested something along these lines (CR 126–127) and the case has been made much more strongly in recent work by Matjaž Vesel (2014). A second influential argument suggests that Copernicus first considered a Tychonic arrangement of the planets and then realized that it transformed into a pure heliocentric system, which again was more systematic and unified (Swerdlow 1973, 434; 2000, 163).

The problem with both of the preceding arguments is that they do not really explain why Copernicanism had to wait until the fifteenth century. The intellectual resources allocated to Copernicus in these arguments were available to anyone who wanted to use them, from antiquity onwards. Admittedly there was a new wave of interest in Plato in the West after the fall of Constantinople, and Copernicus was surrounded by ardent Platonists in Cracow even before he went to Italy for his higher education. And as far as the Tychonic transform argument is concerned, this may be seen as an extension of a theorem in Regiomontanus's *Epitome of the Almagest*, which was published just as Copernicus reached Italy. However, although these possible contributory causes help explain why Copernicus became a heliocentrist, they do not explain why any astronomically proficient Platonists before him, in the Islamicate world or in Europe, failed to use the same resources to develop the same theory.

One remaining explanation, recently suggested by Robert Westman, points to contemporary events as a motivation (Westman 2011; Barker 2013a). First, it is important to understand that, despite the disdain of twentieth-century philosophers, astrology was accepted as a science throughout the early modern period. Indeed the main motive for studying astronomy was to understand astrology, and with the spread of the new technology of printing, many astronomers wrote annual prognostications (predictions of weather, health and politics for the coming year) to make money. Domenico de Novara (1454–1504) did this for the city of Bologna. While Copernicus was studying in Bologna he lived with de Novara and assisted him – at exactly the same time and place astrology was under intellectual attack in the books by Giovanni Pico della Mirandola (1463–1494). Among many complaints Pico mocked astronomers for not being able to establish with certainty the order of the planets. But this is one of the chief results and benefits of adopting a heliocentric system. So, although exposure to Regiomontanus's *Epitome* may have prepared Copernicus to transform geocentric models to heliocentric ones, and his prior commitment to Platonism may have made him value the mathematical harmonies provided by heliocentrism, it may have been the contemporary attack on astronomy that was the proximate cause of his investigating and adopting the heliocentric system (Westman 2011 chapter 3; 2017, 28ff.). It must be said that Copernicus did not make this justification explicit, although he did practice astrology throughout his life and Georg Rheticus made the astrological implications of heliocentrism a centerpiece of the first printed exposition of Copernicus's system (Rheticus/Rosen 2004, 121–127).

We are now a long way from Kuhn's Copernicus, but it is worth summarizing just how non-modern Copernicus was. Although he avoided using the equant, in other respects his system is plainly Ptolemaic. He continued to use eccentrics and epicycles and explicitly accepted the principle that all celestial motions are to be circular or compounded of circles moving at constant speed, although now in the more sophisticated arrangements of Tusi and Urdi. Recognizing the problems originally pointed out by Ibn al-Haytham, he follows the Maragha astronomers and their followers, especially Ibn al-Shatir, in attempting to introduce their solutions to European astronomy. Like them, and like almost all his European contemporaries, he accepts that solid celestial orbs move the planets physically. However, even in his own terms his system has additional problems. There are the problems concerning which orbs rotate "naturally" and which do not (the orb of fixed stars and the Sun are conspicuous exceptions). There is the problem of the gaps between planetary orb clusters. There is the odd treatment of the Earth, which is not carried by an eccentric carrying an epicycle, like the outer planets, but follows a circular path. And perhaps most embarrassing, the center of that orb, or path, and indeed the orbs of the other planets, is a constructed point – the mean sun – not the Sun itself.

It was Kepler who swept all this away and established the modern concept of the heliocentric cosmos. Immediately contradicting Copernicus he revived the equant as a temporary expedient on the way to the ellipse (Kepler/Donahue 2015, chapter 4). Using this stopgap he demonstrated that the planes of all the planets' orbits intersected in the body of the Sun, making his system genuinely heliocentric (Kepler/Donahue 2015, chapter 6). He abandoned epicycles and eccentrics for ellipses and at the same time abandoned Plato's principle that celestial motion must be circular and at constant speed. He abandoned solid celestial spheres for a continuous celestial fluid, in which the planets moved under the influence of a central solar force (although with a bit of animistic self motion too). This enabled him to eliminate Copernicus's third motion, which had only been needed to correct for the effects of the solid orb carrying the Earth. In his system we recognize almost all elements of the modern view of planets moving freely through space, along elliptical paths with the Sun at one focus. If there is an incommensurability in astronomy and cosmology it is between Kepler's system and its Newtonian successors, which are incommensurable with everything that came before, including Ptolemy, the Maragha astronomers and Copernicus (Andersen, Barker and Chen, 2006, 6.5 and 6.6). If Kuhn was writing today, would his book have to be called *The Keplerian Revolution*?

8.5 The Copernican Crisis State

Despite the problems with placing Copernicus historically as ancient – and indeed Islamic – or modern, much of what Kuhn described in his cyclical model does capture the situation from Copernicus to the general acceptance of Newton. Copernicus introduced a novel cosmic scheme with an associated astronomy. The response was not to argue the merits of Copernicus versus Ptolemy and Aristotle. As Paul Feyerabend would have said, there was an efflorescence of new cosmic schemes. By far the most important, and durable, was Tycho's system, which was quickly responsible for the demise of solid celestial orbs and lasted well into the seventeenth century all around the world. For example, Tycho's work was assimilated in China where it led to new instruments (Chapman 1984), star catalogues (Zhang 2008) and cosmic schemes (Deng 2011). But, returning to Europe, we should also note many variations on Tycho, Copernicus and, indeed, updated Aristotelian systems.

Tycho's system was adopted widely in both its original and in modified forms, although Tycho was not always identified as the author. Perhaps his authorship was denied or passed over because of competing claims or perhaps because he was a Lutheran and many who adopted his system were Catholics. Erycius Puteanus (1594–1646) was a celebrated linguist who taught for many years at the Catholic university of Louvain. Johann Baptista Cysatus (1586–1657) was a Swiss Jesuit. Both adopted the Tychonic system without crediting Tycho although Cysatus even adopted Tycho's theory of the motion of comets (Schofield 1981, 170–171).

The most notorious competitor for authorship of the geo-heliocentric system was Nicholaus Reimers Baer (Latin: Ursus, 1551–1600), who served as court mathematician to the emperor Rudolf II until he was replaced by Tycho Brahe himself. In 1588, the same year that Tycho announced his system, Ursus published an alternative that differed in two important respects (Ursus 1588; Ursus/Thierfelder & Launert 2012). First, although Ursus also denied the existence of solid celestial orbs carrying the planets, the path of the planet Mars was enlarged so that it no longer crossed the path of the Sun. Second, Ursus attributed a daily rotation to the Earth (Schofield 1981, 108–119). Tycho bitterly attacked Ursus for plagiarism (Jardine and Segonds 2008), even commissioning Kepler to write a lengthy rebuttal of Ursus's work (Jardine 1984). However, for our purposes the differences between the two systems are quite sufficient to count them as distinct cosmic schemes (Launert 2009).

A third option was proposed by Helisaeus Roeslin (1545–1616), an astrologer and court physician in Alsace, who corresponded with many of his contemporaries. After meeting Ursus, but apparently in ignorance of Tycho's work, in 1597 Roeslin published a book remedying what he saw as the defects of Ursus's system and explicitly comparing his new proposal to the systems of Ptolemy, Copernicus, Ursus and Tycho (Roeslin/Granada 2000, 43–56; Launert 2009). He proposed to eliminate the rotation of the Earth, and reintroduced solid orbs to transport the planets. The Mars-Sun collision problem was avoided by expanding the orbs of all three outer planets (Schofield 1981, 136–144). Although the latter suggestion made his scheme problematic astronomically, it was widely read and, again, clearly counts as a distinct cosmic scheme.

In Paris in 1623 Jacques du Chevreul (1595–1649) published an Aristotelian cosmic scheme that accommodates all Galileo's telescopic discoveries (Ariew 2006). He achieved this by embracing Galileo's discovery of the moons of Jupiter and making primaries surrounded by satellites the main pattern in the heavens. Thus, Saturn has two satellites (following Galileo), Jupiter has four, and the Sun has two: Mercury and Venus. Only Mars and the Moon circle the Earth without companions. His system remains geocentric and geostatic. Du Chevreul was an influential teacher at the university of Paris. The book in which he proposed this renovation of Aristotle's cosmos went through sixteen editions in three languages between 1623 and 1649 (WorldCat 2020).

As late as 1651 careful weighing of the scientific evidence still favored a Tychonic world system (Graney 2015; Dinis 2017). In the *New Almagest*, Giovanni Battista Riccioli (1598–1671) examined 126 arguments for and against Copernicanism. Many he dismissed as trivial, but the nontrivial ones amounted to a clear case against Copernicus and, as Ptolemy and

Aristotle were no longer an option, in favor of something like Tycho's system. Especially weighty were the experimental failure to detect the rotation of the Earth and astronomical arguments from the sizes and distances of stars (Graney 2015, 115–139; Dinis 2017). Careful Jesuit experiments on falling bodies had also undermined Galileo's claims (Graney 2015, 87–101). Neither Galileo's nor Copernicus's work looked scientifically sound. Emphatically setting aside arguments from the Bible or the authority of the Church, Riccioli concluded: "[T]he hypothesis that assumes the motion of the Earth, be it only daily or both daily and annual, is to be declared false and repugnant on the basis not only of physical but also of physico-mathematical demonstrations" (Riccioli, 1651, II: 478, tr. Dinis 2017, 287).

Even on the Copernican side, many major figures offered new cosmic schemes. As we have already seen, Kepler's system differs so much from Copernicus's that it could have been presented as completely novel. Galileo rejects Kepler's elliptical orbits and the orb of fixed stars accepted by both Copernicus and Kepler as the boundary of the cosmos. Hence his view counts as a separate scheme. And Descartes's celebrated cosmos of material vortices differs from all of these. The continuing credibility of several options, and the difficulty of choosing between them, is illustrated by the refusal of well-informed scholars like John Milton (1608–1674) to take sides. In *Paradise Lost* (1667) he shows an awareness of all the major options but does not settle on any one of them (Danielson 2014).

It was only the acceptance of Newton's synthesis, built on Kepler and to a lesser extent on Galileo, that brought this period to its end. In the West, the period from Copernicus to Newton was a crisis state with no single generally accepted paradigm or normal science tradition, and the emergence of multiple competing alternatives, one of which starting with Kepler, and leading through Newton, became the new paradigm. This process was slow and in some respects cumulative. Major changes were initiated by the proposal of heliocentrism, the abandonment of solid celestial orbs and the replacement of purely circular celestial motions. However, by the end of it (and perhaps as I have suggested earlier, by the middle of it) there was significant incommensurability with the previous paradigm, although this seems to have caused little in the way of incomprehension or communication failure. Ironically the new system that suffered most from this seems to have been Tycho's, which was wildly misunderstood by people who could not quite see past solid orbs. Copernicus's work was not revolutionary, but the crisis state it caused did destroy the old paradigm and finally produce a new one.

CHAPTER 9

Kuhn, the Duck, and the Rabbit – Perception, Theory-Ladenness, and Creativity in Science

Vasso Kindi

9.1 Introduction

Kuhn, in his letter dated November 29, 1962, explained to Edwin B. Boring, a professor of psychology at Harvard University, why he chose the duck-rabbit figure to discuss revolutions and theory-ladenness in SSR. Boring had suggested that Kuhn use the goblets and the young woman/old woman drawings as they are more complicated than the duck-rabbit. Here is Kuhn's reply: "The duck-rabbit has by now become almost a cliché in a great many circles, particularly since Wittgenstein discussed it at such length. That is why ... I selected the duck-rabbit." Kuhn is referring to Wittgenstein's discussion of the duck-rabbit figure in the context of Wittgenstein's remarks on "seeing" and "seeing as" in what is now called "Philosophy of Psychology – A Fragment, Section xi" (1953/2009 PPF, §118).[1] Let us see first what Kuhn did with the duck-rabbit in SSR and how he profited from Wittgenstein's discussion of it. I will then examine the implications of the use of the duck-rabbit metaphor by Kuhn in his work and I will close with an analysis of Kuhn's account of creativity. My key contention is that Kuhn, influenced by Wittgenstein, rejects a two-tier account of perception, that is, seeing raw data first and interpreting them later, and does not assimilate all "seeing" to "seeing as". The use of the duck-rabbit figure helps Kuhn elucidate how he understands scientific revolutions and world changes; it helps him make the logical point that radical transformation has to be holistic and not piecemeal. It also helps him account for creativity since Kuhn associates novelty with the change of aspect.

---

[1] In the first three editions of Wittgenstein's *Philosophical Investigations*, the book was divided into two parts: Part I and Part II. In the fourth edition, the editors gave Part II the title "Philosophy of Psychology – A Fragment" and numbered the remarks. I will cite from the fourth edition.

9.2 Kuhn's Use of the Duck-Rabbit

Kuhn makes use of the duck-rabbit metaphor for the first time in Chapter X of SSR, on revolutions as changes of world view.[2] He says that "what were ducks in the scientist's world before the revolution are rabbits afterwards" (SSR-4, 111–112). The next sentence, "The man who first saw the exterior of the box from above later sees its interior from below" (SSR-4, 112), alludes to Wittgenstein's discussion of a schematic drawing that can be seen as an upturned open box (1953/2009 PPF, §116).[3] So we have a clear indication that Kuhn's use of the gestalt figures draws on Wittgenstein's discussion of them. We will consider later what he borrows from Wittgenstein. For now, let us see what use Kuhn makes of these visual gestalts.

Kuhn says that the gestalt figures are "suggestive" of what goes on in the scientist's world after a revolution (SSR-4, 111). What do they suggest? That the scientist's worlds before and after a revolution resemble the two different aspects, duck and rabbit, of the gestalt figure. They also suggest that the paradox associated with these figures, namely that the stimulus remains unaltered even though the perception of it changes, also characterizes the perception of scientists. One might say that, in the case of the duck-rabbit, what remains the same are the lines on the paper which may be perceived differently by the subjects who look at them, while in the case of the scientists, what remains unaltered is a "hypothetical fixed nature" (SSR-4, 118) that is perceived differently by them. As Kuhn put it, "though the world does not change with a change of paradigm, the scientist afterward works in a different world ... What occurs during a scientific revolution is not fully reducible to a reinterpretation of individual and stable data" (SSR-4, 121).

Kuhn notes that, in order to see just the lines without seeing the figures, one needs to *learn* to concentrate on them. In these particular circumstances, "[the scientist] may then say (what he could not legitimately have said earlier) that it is these lines that he really sees but that he sees them

---

[2] Earlier in the book (SSR-4, 85), Kuhn made a reference to visual gestalts used by Hanson (1958), namely to the ambiguous figure that can be seen as either a bird or an antelope.
[3] Joseph Weiss, a psychoanalyst, who was Thomas Kuhn's friend when they were both in California in the late 1950s and 1960s, remembers in an interview (Andresen 1999, S65) that Kuhn was "very interested" in the Necker Cube, a schematic drawing of a cube that can be seen in two different ways: as looking downwards and as looking upwards. The Necker cube was also of interest to Wittgenstein in the *Tractatus* (5.5423), where he said that looking at this drawing we perceive two different facts, and in the *Philosophical Investigations* where it is not mentioned by name but it is discussed as a schematic cube that has various aspects (e.g., Wittgenstein 1953/2009 PPF, 135).

alternately *as* a duck and *as* a rabbit" (SSR-4, 113). Kuhn's point is that, normally, one sees either the duck or the rabbit and, only in case one has learned to concentrate on just the lines, can one say that they see the lines as something else, that is, *as* a duck or *as* a rabbit. Kuhn, that is, preserves the use of "seeing as" for special occasions. This is in contradistinction to Hanson who approvingly cites the British philosopher G. N. A. Vesey who said that "all seeing is seeing as ... if a person sees something at all it must look like something to him" (Hanson 1958, 182). Vesey's statement appears in a paper that he delivered in 1956 at a Meeting of the Aristotelian Society entitled "Seeing and Seeing As". In that paper Vesey investigates, in the spirit of ordinary language philosophy, the use and meaning of expressions such as "It looks like a torpedo, a lemon, etc." and "It is a torpedo, a lemon, etc." In that context, he also discusses the expression "I see x as ... " and says that "if a person sees something at all it must look like something to him, even if it only looks like 'somebody doing something'" (1956, 114). Vesey is interested in whether a judgement is involved in perception and his conclusion is that it is not necessarily involved although "experience and judgment are connected: for what an object looks like to a person is what he would judge that object to be if he had no reason to judge otherwise" (ibid., 124). How a thing looks, he says, "is something phenomenal, not intellectual" (ibid). In the same article, Vesey refers to the reversible figures in psychology and says that they are used in textbooks to indicate that perception "is a function of the receiving organism as well as of the stimulus ... Perception is a function of both stimulus and receptor throughout" (ibid., 112).

Hanson may approvingly cite Vesey's "all seeing is seeing as" but he is aware that Wittgenstein, on whose work he relies,[4] does not share this conviction. He admits that Wittgenstein is reluctant to concede the identification of *seeing* with *seeing as* and claims that he does not understand Wittgenstein's reasons (1958, 19). Hanson does not take seeing to be a two-stage process[5] – "Seeing an X-ray tube is not seeing a glass-and-metal

---

[4] On Wittgenstein's influence on Hanson, see Kindi (2016, 594–595).
[5] Hanson (1958, 16–17) cites from Pierre Duhem's *La théorie physique* in order to combat the view that the physicist and the layman see the same thing but interpret it differently. The passage he chooses describes what a layman and a physicist see in a laboratory. The layman sees spools and an oscillating iron bar carrying a mirror, and the physicist sees the measuring of the electric resistance of spools. Hanson says that "the visitor must learn some physics before he can see what the physicist sees" (ibid., 17). He also asks whether the physicist is doing more than just seeing, and he answers: "No; he does nothing over and above what the layman does when he sees an X-ray tube" (ibid., 16). So Hanson marshals Duhem as an ally for his thesis that observation is theory-laden and does not involve interpreting bare facts. The problem is that Duhem may not be the right ally. Right after the passage cited by Hanson, Duhem says that any experiment in physics involves two parts: the first part is the observation of facts and the second part is the interpretation of the observed facts (Duhem 1982/1914,

object as an X-ray tube". But he still insists that "the logic of 'seeing as' seems to illuminate the *general* perceptual case" (ibid., my emphasis). It is at this point that Hanson cites Vesey. Wittgenstein does not take seeing to be a two-stage process either but, unlike Vesey or Hanson, does not think that "seeing as" illuminates the general perceptual case. Kuhn was influenced by Hanson (SSR-4, 113; RSS, 311), but on this point he is closer to Wittgenstein. Kuhn says that "scientists do not see something *as* something else; instead, they simply see it" (SSR-4, 85).[6]

9.3 Wittgenstein on *Seeing* and *Seeing As*

In Wittgenstein's view *seeing as* "is not part of perception. And therefore it is like seeing, and again not like seeing" (1953/2009 PPF, §137). I am reporting my perception if, shown the duck-rabbit picture and asked what it is, I reply "It's a rabbit", "It's a duck" or "It's a duck-rabbit" (1953/2009 PPF, §128). According to Wittgenstein, in all these answers I describe my perception, although in the first two cases, the ambiguity of the picture escapes me. In the case of "seeing as", however, although we may find things in favour of saying that "seeing it as" is a sensation, "we have to describe this sensation as though we were describing an interpretation" (Wittgenstein 1988, 332).[7] Giving an interpretation involves a two-stage process: first, seeing dashes, strokes, shapes, colours and the like, and then making some kind of a hypothesis (1953/2009 PPF, 249) or a comparison to a paradigm, "as if something had been pressed into

---

145). One may surmise that Duhem is in agreement with Hanson on the theory-ladenness of observation (the observation of facts), reserving theoretical interpretation, as an extra process, for what comes after observation. Unfortunately, this supposition does not seem to be correct. Duhem believes that observation provides us with concrete data that are later replaced by "abstract and symbolic representations" provided by interpretation (ibid., 147). He says that it is not necessary to know physics to observe facts; knowledge of physics is necessary only to interpret facts (ibid., 145). He also says that "an experiment in physics is the precise observation of phenomena accompanied by an interpretation of these phenomena" (ibid., 147). In general, Duhem thinks that theoretical interpretation is essential in the case of physics, but not in the case, for instance, of physiology where it is possible to leave theory outside the laboratory door. In an experiment in physics, he says, "[W]ithout theory it is impossible ... to interpret a single reading" (ibid., 182). So it seems that for Duhem, observation does not need theory and may be common between laypersons and scientists. Theory is needed in order to proceed to interpretation which is, according to Duhem, indispensable for physics.

[6] Cf. what Joseph Weiss reports about what Kuhn believed in relation to gestalt figures such as the Necker cube: "that a person, inexplicably, just saw things a certain way – one just gets it ... one just sees it" (Andresen 1999, S65).

[7] Cf. "The puzzle is that 'seeing as' is described on the paradigm of interpretation" (Wittgenstein 1988, 102).

a mould it did not really fit into" (1953/2009 PPF, §164; cf. 1958, 168–169). It is an indirect description (1953/2009 PPF, §117).

Wittgenstein is reluctant to assume that, in the case of ambiguous figures, we interpret differently something common and more basic (1953/2009 PPF, §248). He connects *seeing* and *seeing as* to the broader issue of meaning and experience of meaning. As in the case of seeing, where, without thinking, we immediately perceive something, we immediately perceive meaningful expressions, linguistic or other (for instance, gestures). We do not separately perceive dead, inorganic, signs that later become alive when imbued with meaning (1958, 4). The individuation of the sign itself requires that we treat it as a symbol, that is, as significant (cf. Conant 2020). Seeing an aspect, or seeing X *as* Y, on the other hand, has "a close kinship with 'experiencing the meaning of a word'" (1953/2009 PPF, §234; cf. §261). We experience meaning when words, even if unaltered, acquire "a different ring" (1953/2009 PPF, §264), when we read them with an intonation or with a certain feeling (ibid.). For Wittgenstein, seeing aspects is parasitic upon regular seeing, and experiencing the meaning of words is parasitic upon the common use of words (cf. Cavell 1979, 355; Kindi 2009). Unlike seeing which is ubiquitous, "seeing as" "does not come in ordinary life" (Wittgenstein 1988, 231). We would not say, at the sight of a bottle of wine, "Now I'm seeing this as a bottle" (1982/1990, §534); nor would we normally say it of a knife and fork (1982/1990, §535).

> It would have made as little sense for me to say "Now I see it as . . ." as to say at the sight of a knife and fork "Now I see this as a knife and fork". This utterance would not be understood. Any more than: "Now it is a fork for me" or "It can be a fork too" (1953/2009 PPF, §122).
>
> One doesn't "*take*" what one knows to be the cutlery at a meal *for* cutlery, any more than one ordinarily tries to move one's mouth as one eats, or strives to move it. (1953/2009 PPF, §123)[8]

Wittgenstein is not against interpretation *tout court*. We may interpret words, sentences and texts. Think of what literary critics and philologists do. But this kind of interpretation is a process that takes time, involves hypotheses and depends upon previous immediate recognition and understanding of linguistic items. It is not an activity that breathes meanings into

---

[8] Here Wittgenstein is alluding to his view that *seeing as* is subject to the will: "Seeing an aspect and imagining are subject to the will. There is such an order as 'Imagine *this*!', and also, 'Now see the figure like *this*!'; but not 'Now see this leaf green!'" (Wittgenstein 1953/2009 PPF, 256).

dead signs. It already deals with meaningful units. As Wittgenstein said, "interpreting a sign, adding an interpretation to it, is a process that does take place in some cases but certainly not every time I understand a sign" (Wittgenstein 2005, 16; cf. 1967/1970, §218). The fact that we engage in interpreting texts, or that on certain occasions we need to interpret particular words or phrases, does not mean that an interpretation is a prerequisite to understanding. In the ordinary use of language, we do not interpret physical marks; we simply grasp what they say. "Every sign is capable of interpretation; but the *meaning* mustn't be capable of interpretation. It is the last interpretation" (Wittgenstein 1958, 35). The same holds for seeing: if someone threatens me with a knife, I do not have to add an interpretation to what I immediately perceive; nor do I need to make inferences. "What if I were to say: It isn't enough for me to perceive a threatening face – first I have to interpret it. – Someone pulls a knife and I say: 'I understand this as a threat'" (Wittgenstein 2005, 16).

9.4 The Rejection of the Raw Data + Interpretation Understanding of Perception

Kuhn follows closely Wittgenstein's take on interpretation. In discussing the transition from one paradigm to the next, he repeatedly argues that scientists do not share raw observations that they then interpret differently. The data scientists receive, Kuhn claims, are not jointly fixed by their physiology and the environment alone; their immediate experience is permeated by the paradigm they have been trained with. Galileo and Aristotle did not see the same thing looking at falling stones. One saw a pendulum, the other saw constrained fall (SSR-4, 125). The scientists' most elementary experiences are seeing pendulums, planets, oxygen, atoms and electrons. In principle, they cannot have more elementary experiences than these (SSR-4, 127–128). They do not see lines, colour patches and the like *as* pendulums, planets, oxygen, atoms and electrons. Even retinal imprints that are supposed to be captured by a neutral observation-language are, according to Kuhn, "elaborate constructs" (SSR-4, 127).[9] Equally constructed, and not simply "given", are operations and measurements in scientific laboratories. The collection of this kind of data requires much effort and expertise and presupposes a paradigm (SSR-4, 125–126).

---

[9] Kuhn attributes the idea of a pure observational basis that subsequently gets interpreted to an epistemological paradigm associated with Descartes (SSR-4, 121, 126). This paradigm, Kuhn says, has not succeeded in yielding a language of pure percepts. "Philosophical investigation has not yet provided even a hint of what a language able to do that would be like" (SSR-4, 127; cf. SSR-4, 96).

This means that there is no fixed and neutral perceptual basis that is interpreted in different ways. Furthermore, in Kuhn's view, experience is fluid and it does not lend itself to a piecemeal process that imposes an interpretation on isolated articles of perception. That is the reason why the change of paradigms brings about a transformation of the whole bundle of experiences (SSR-4, 123).

The rejection of this two-tier picture of knowledge and perception (raw data + interpretation) does not impel Kuhn, as it did not impel Wittgenstein, to reject interpretation altogether. Kuhn acknowledges that scientists engage in interpreting data. He even says that this task is central to the scientific enterprise. But he warns that these interpretations take place under the guidance of a paradigm and are only carried out in the process of articulating it in the practice of normal science (SSR-4, 122). They do not induce a change of paradigms by supposedly offering the correct appraisal of bare percepts. With the change of paradigms the world changes along with it. There is no recourse to a fixed nature that is interpreted differently.

This was a very provocative claim. Kuhn made it with great caution and many qualifications. But his critics, even his allies, were nevertheless shocked. How could this be? Hempel (1980, 197) expressed this concern humorously: "If adherents of different paradigms did inhabit totally separate worlds, I feel tempted to ask, how can they ever have lunch together and discuss each other's views?" Let us look at the issue more closely.

9.5 World Changes

In the case of ambiguous figures, such as the duck-rabbit, observers know that they are looking at the same drawing while their perception of the figure shifts. They may be holding the book with the drawing in their hands, or they may have been told by an authority that they are looking at an ambiguous figure. This is the reason they can say that they see the figure alternately *as* a duck and *as* a rabbit. If, for whatever reason, the ambiguity of the figure escaped them and they had nobody to tell them that they were, for instance, aspect-blind, then they would not notice a change of aspects; they would simply see a duck or a rabbit while looking at the drawing. In other psychological experiments, such as the one with the anomalous playing cards conducted by Jerome Bruner and Leo Postman and discussed by Kuhn in SSR (SSR-4, 62–64), the stability of the stimulus is guaranteed by the experimenter who acts as the authority the subjects of the experiment can resort to when they realize that their perception shifted.[10] In the case of paradigm shifts, however, the scientists do not have recourse to a fixed nature that they come to see differently, nor can they be assured by an authority that they are looking at the same world when their perception changes (SSR-4, 118). According to Kuhn, their most elementary perceptions are shaped by different paradigms and, so, scientists cannot have independent access to a bare, neutral, unmediated world. With the change of paradigm, the world changes.

Now, Kuhn is very guarded in making this claim. He says at the beginning that "the historian of science *may be tempted to* exclaim that when paradigms change, the world itself changes with them" (SSR-4, 111, emphasis added). He later says that it is "*as if* the professional community had been suddenly transported to another planet" and that "*we may want to say* that after a revolution scientists are responding to a different world" (ibid., emphases added). He clearly distances himself from categorically asserting these statements. He is not endorsing them; he entertains them to see where they would lead. At the same time, Kuhn is also eager to emphatically declare that scientists are not transported to another planet: "[N]othing of quite that sort does occur: there is no geographical transplantation; outside the laboratory everyday affairs usually continue as before" (ibid). And later he contends that the changes in tests and measurements after a revolution "are never total"; "the scientist after a revolution is still looking at the same world"; "much of his language and most of his laboratory instruments are still the same as they were before" (SSR-4, 129). Still, Kuhn insists that "we must learn to make sense of statements that at least resemble these", that is, statements such as this one: "[T]hough the world does not change with a change of paradigm, the scientist afterward works in a different world" (SSR-4, 121). This statement is strictly speaking a contradiction. How should we make sense of that?

Paul Hoyningen-Huene (1993, 31–63) tried to make sense by distinguishing between two worlds: the noumenal world ("the world-in-itself") that does not change with the revolution and the phenomenal world that does change with it. With this distinction the statement is not

---

[10] The subjects of Bruner and Postman's experiment were shown anomalous playing cards, such as a black four of hearts, and they were asked to identify them. When the exposure to the cards was short, the subjects of the experiment tended to assimilate what they saw to already familiar categories; for instance they would call the card either a black four of spades or a red four of hearts. As the exposure increased, they began to realize the problem and, finally, most of them succeeded in identifying the cards as anomalous. They realized that their perception had changed because the experimenter assured them that all along they were looking at the same cards.

contradictory anymore. Rephrased, the statement would read as follows: though the noumenal world does not change with a change of paradigm, the scientist afterward works in a different phenomenal world. This is a philosophical solution to the problem. By devising, or following a philosophical theory, Kant's in this case, the problem disappears. To be sure, Hoyningen-Huene did not simply come up with this theory to fix the problem. He traced Kuhn's use of the relevant terms and found confirmatory evidence that allowed him to attribute the theory to Kuhn.

Kuhn, however, did not want to simply get rid of the problem. On the contrary, he wanted to highlight the difficulty of giving up what he called the old Cartesian epistemological paradigm, which has shaped both our understanding of things and the language we speak for centuries and at the same time embrace an approach that had yet to be formed.[11] As he put it, "*In a sense that I am unable to explicate further*, the proponents of competing paradigms practice their trades in different worlds" (SSR-4, 149, emphasis added). Kuhn is at a loss for words: he has come to see that the perception of scientists is never that of pure data, as the received epistemological paradigm prescribed, and tries with great difficulty to accommodate this new insight within the received framework. So Kuhn would not be happy, I think, to make the problem disappear; his aim was to deepen our appreciation of the problem. Besides, as Hoyningen-Huene correctly acknowledges (Hoyningen-Huene 1993, 60), Kuhn himself explicitly rejects the Kantian world-in-itself. "The view toward which I grope would also be Kantian, but without 'things in themselves' and with categories of the mind which could change with time"[12] (RSS, 207). Although Kuhn again expresses his bafflement by saying that he *gropes* to formulate his view, it is clear that his Kantianism is restricted to acknowledging the contribution of the categories of the mind, even if moveable, to

---

[11] "In the absence of a developed alternative, I find it impossible to relinquish entirely that viewpoint [the epistemological viewpoint that understands theories as interpretations of data]" (SSR-4, 125).

[12] It should be noted, however, that in his paper "The Road since *Structure*", delivered in 1990 as a presidential address to the biennial meeting of the Philosophy of Science Association, and I suppose in response to relentless criticism that he is endorsing different kinds of idealism and to his own worries that he may be sliding in that direction, he recognizes the need for "something permanent, fixed, and stable" to serve as a source of stability in view of all differentiation and change (RSS, 104). He compares it to Kant's *Ding an sich* and says that, as such, "it is ineffable, undescribable, undiscussible" (ibid). He even gives it an un-Kantian genetic dimension saying that it is "the whole from which have been fabricated both creatures and their niches, both the 'internal' and the 'external' worlds" (ibid.). As the editors of RSS observe in their introduction, Kuhn wavered on the Kantian "Ding an sich". It seems, though, that his considered position, intimated to the same editors before his death, was to repudiate "both that notion and the reasons he had put forward for it" (RSS, 7).

knowledge and experience. He finds the idea that there is a world out there that science is zeroing in on meaningless (RSS, 243). The positing of such an unknown and unknowable "world-in-itself" seems only to serve to appease our ontological worries. Kuhn wants to be a realist (RSS, 206). But he is not ready to concede realism's emblematic claim. His world is not an external, mind-independent world plain and bare, but one that is populated by the entities science approves.[13] But what science approves changes, not simply by accumulation, that is, by discovering more entities or learning more about them. Successive scientific theories transform our conception of the world. And our conception of the world is not just an interpretation of a given reality (RSS, 95). Kuhn's world is one and pliable. He finds the contrast between mind-independent and mind-dependent misleading and opts for the view that "it is groups and group practices that constitute worlds (and are constituted by them)"[14] (RSS, 103). That is why he speaks of "cultural ontology" (RSS, 246) and shows such an interest in niches, that is, the habitats that creatures and environments build together in interaction. "A niche is the world of the group which inhabits it, thus constituting it a niche" (RSS, 103; cf. 120, 250). Kuhn is struggling with language, with the metaphors it forms, and with the distinctions it makes in order to express his new way of seeing things. What he tries to say turns out to be contradictory or ungrammatical or simply contrary to what we expect or are used to hear. But this is how things are when a revolution is in the offing. Kuhn described the phenomenon in relation to science: "Violation or distortion of a previously unproblematic language is the touchstone of revolutionary change" (RSS, 31). Kuhn is not trying to revolutionize science, but philosophy of science. He finds it "impossible to relinquish entirely" the old epistemological paradigm traced back to Descartes (SSR-4, 125), which pits language against a mind-independent, external world and takes theories to be interpretations of given data. And he is eager, at the same time, to embrace and express a new revisionary alternative that is still not yet developed.

Let us take stock. Kuhn used the duck-rabbit figure as a metaphor for revolutionary change in science. The two aspects of the drawing stand for the two ways of perceiving the world, before and after a revolution. The drawing itself stands for the common stimulus that is seen differently or for the common world that is transformed. The two aspects are not two interpretations of common raw data and perception is not normally

---

[13] "What is the world, I ask, if it does not include most of the sorts of things to which the *actual* language spoken at a given time refers?" (RSS, 206).
[14] Cf. "In much of language learning these two sorts of knowledge – knowledge of words and knowledge of nature – are acquired together, not really two sorts of knowledge at all, but two faces of the single coinage that a language provides" (RSS, 31).

a matter of seeing as. People who observe ambiguous figures and scientists who have been trained under different paradigms collect different data (SSR-4, 121). Kuhn was influenced by both Hanson and Wittgenstein in his views on seeing. All three of them rejected the view that perception is a two-stage process and were more inclined to hold that observation is theory or concept-laden[15]. Kuhn, however, siding with Wittgenstein, differed from Hanson in rejecting the claim that all *seeing* is *seeing as*, despite the assimilation of their views by critics.[16]

Now, it was mentioned earlier that the duck-rabbit metaphor was *only suggestive* of what goes on in science in revolutionary periods (SSR-4, 111).[17] This means that the analogy is not perfect. Apart from the similarities that were already discussed, there are also differences. These differences were highlighted by Kuhn's critics, in order to avoid the undesirable consequences of the analogy for science, but were anticipated by Kuhn himself already in SSR. For instance, Kuhn noted that, in scientific change, there is no external authority to assure the scientists that they are dealing with the same world and the same stimuli when they come to see things differently. This is the reason that scientists do not see things *as* X or Y but simply see them (SSR-4, 85), By contrast, the subjects of the psychological experiments, who either know that they are looking at ambiguous figures or are assured that they are looking at the same cards despite their different perceptions, see the figures *as* ducks or rabbits and the cards *as* anomalous or not. Another difference is that the scientists cannot switch back and forth as the subjects of the gestalt experiments can (SSR-4, 114–115). Once they embrace the new paradigm, they usually do not go back.[18] Also, scientists only rarely give voice to their experience of shifting vision, as it

---

[15] To be sure, Wittgenstein never used such expressions. But if we try to translate Wittgenstein's thoughts as regards perception into Hanson's terminology, we may be allowed to say that he was in favour of concept-laden observation, given that, in his view, in normal circumstances, we immediately make sense of what we observe – we do not apprehend unintelligible marks or colour patches that we later put an interpretation on. Cf. "Here we must be careful not to think in traditional psychological categories. Such as simply dividing experience into seeing and thinking; or doing anything like that" (Wittgenstein 1982/1990, § 542).

[16] Raftopoulos (2009, 312), for example, says that both Hanson and Kuhn "rendered the distinction between *seeing* and *seeing as* obsolete." That is certainly not true of Kuhn.

[17] "Though psychological experiments are suggestive, they cannot, in the nature of the case, be more than that" (SSR-4, 113–14). Cf. "We need not insist on so full a parallelism" (SSR-4, 117).

[18] One may think, however, that "bilingual" scientists, e.g., scientists familiar with two consecutive paradigms, the one they were originally trained with and the one they eventually endorse, would be able to alternate and to converse in both languages. Yet the idea, I think, is that once a paradigm is adopted in the sciences, all previous ones become obsolete. As Cavell (1979, 371) put it: "Once convinced of Continental drift, there is no competing picture of the formation of continents to which one is liable to revert. One sees here something of what scientific progress means."

may take time to acquire a new way of perceiving things. This means that the experience of instantaneous conversion is also rare in science and, therefore, unlike what goes on with the gestalt figures. So, despite sporadic references to Pauline experiences of illumination, Kuhn does not think that the scientific community converts to the new way of seeing collectively, as a group. What happens, in his view, is "an increasing shift in the distribution of professional allegiances" (SSR-4, 157).[19]

Apart from the similarities and differences that have already been noted between the ambiguous figures and radical change in science, the duck-rabbit metaphor is also thought to be suggestive in another sense: It hints at similarities between the so-called lighting up of an aspect in the case of ambiguous figures and the advent of novelty. Just as one sees something new and unexpected in aspect perception, scientists break new ground and become creative by seeing things in the world differently. In the last part of this chapter, I will show the implications this comparison had on Kuhn's views about creativity and innovative thinking.

9.6 Creativity

Creativity implies novelty, and revolutions in science, together with artistic innovation, are considered prime examples of creative practice.[20] Kuhnian revolutions, in particular, which usher in a new way of seeing the world, are supposed to mark deep ruptures with the way the world was previously experienced. In that sense they are compared to the dawning of an aspect which brings to light a novel way of perceiving the same stimuli. One implication of this comparison is that, for Kuhn, advances in science do not come from without as additions to the already available pile of beliefs, but are the result of reassembling and reconfiguring old material.

> The transition from a paradigm in crisis to a new one from which a new tradition of normal science can emerge is far from a cumulative process, one achieved by an articulation or extension of the old paradigm. Rather it is a reconstruction of the field from new fundamentals ... One perceptive historian, viewing a classic case of a science's reorientation by paradigm change, recently described it as ... a process that involves "handling the same bundle of data as before, but placing them in a new system of

---

[19] In "Reflections on My Critics" (RSS, 123–170), Kuhn elaborates on the differences between the change of perspective in the gestalt figures and conceptual change in science and clarifies his view. Cf. RSS, 56–57.
[20] For a comparison between science and art as regards novelty and revolution in relation to Kuhn, see Kindi (2010).

relations with one another by giving them a different framework." Others who have noted this aspect of scientific advance have emphasized its similarity to a change in visual gestalt: the marks on paper that were first seen as a bird are now seen as an antelope, or vice versa. (SSR-4, 85; cf. ET 226–227)[21]

The other implication of the comparison between the gestalt switch and creativity in science is that there is something inexplicable in the creative process, that is, in the process that induces a change of perspective. The transition from the old to the new is not seen as a piecemeal operation dictated by logic and empirical evidence but as brought about by conversion. As Arthur Koestler (1981/1985, 15) put it in relation to creativity in general, "the creative act itself is for the scientist, as it is for the artist, a leap into the dark".[22] Creativity has certainly been associated, especially in the Romantic model, with the mystical powers of a genius, with intuitionist strokes of illumination and with inspired flashes of insight (Runco & Albert 2010; Nickles 1994). It is misleading, however, to attribute this mystical and irrational understanding of creativity in science to Kuhn, despite his use of the gestalt switch metaphor to account for scientific advances. Kuhn did in fact refer to expressions such as "scales falling from the eyes" and "lightning flash", used by scientists when they suddenly saw a solution to the puzzle that troubled them. And he repeatedly told the story of his own "Aristotle experience", that is, the experience of suddenly making sense of Aristotle's *Physics* which he previously could not understand (ET, xi–xii; RSS, 15–17, 292–293; Sigurðsson 1990/2016, 21). "This sort of experience", he said,

> the pieces suddenly sorting themselves out and coming together in a new way – is the first general characteristic of revolutionary change ... Though scientific revolutions leave much piecemeal mopping up to do, the central change cannot be experienced piecemeal, one step at a time. Instead, it involves some relatively sudden and unstructured transformation in which some part of the flux of experience sorts itself out differently and displays patterns that were not visible before. (ET, 17)

---

[21] The references are to the historian Herbert Butterfield and the philosopher N. R. Hanson who used the bird-antelope figure in his work.

[22] Cf. "There are always large chunks of irrationality embedded in the creative process, not only in art (where we are ready to accept it) but in the exact sciences as well" (Koestler 1985, 14). Cf. MacIntyre (1977) for criticizing Kuhn's revolutions as leaps in the dark and as introducing irrationality in scientific development.

Still, I believe that these references and statements are not supposed to underwrite the irrationality of scientific development. Kuhn was taken aback by this kind of criticism and was totally opposed to the idea: "It is emphatically *not* my view that 'adoption of a new scientific theory is an intuitive or mystical affair'" (RSS, 157).[23] Sudden illumination was not for him a matter of mystical inspiration but the result of extensive practice and familiarization with the problems and the superseded frameworks.[24] The point Kuhn wanted to make by the gestalt switch metaphor and the references to conversion was that scientific advance is not a matter of theoretical proof, but a matter of changing allegiance. Theoretical proof presupposes agreement in the premises and Kuhn thought that this agreement was not there because of, among other things, meaning variance across the revolutionary divide. So, when the line of argument reached deadlock, adherents of different paradigms had to be persuaded to endorse a different way of seeing things. The transformation had to be holistic since scientific frameworks were for Kuhn holistic structures built around hinge exemplars and not sets of logically related independent statements. However, it did not have to be instantaneous. Conversion could take years to be effected by means of persuasive argumentation (SSR-4, 86, 94, 150, 157).[25] As mentioned earlier, the instantaneous gestalt switch was only an analogy that captured some, and surely not all, aspects of revolutionary change.[26] Kuhn used it because he had empirical evidence from historians and scientists who had reported such episodes. But, most importantly, he used it because it served him, on the logical level, as the opposite of a step-by-step proof.[27] If science were to advance by accumulation, making inferences from experience, a gestalt switch would be completely out of place in that picture. In Kuhn's model, however, where transition from one paradigm to the next is not a piecemeal enterprise controlled by experiment and logic, but a reconfiguration of a holistic structure, the metaphor of an instantaneous gestalt switch served him

---

[23] Here Kuhn refers to Israel Scheffler's criticism. Cf. "[T]he notion that I was showing the irrationality of science absolutely blew my mind" (Sigurðsson 1990/2016, 22).
[24] Cf. Blackburn (2014, 151): "But even the anecdotes [about 'aha' experiences and 'eureka' moments] point out that the illumination requires a thoroughly prepared mind."
[25] Kuhn never meant to exclude rational argumentation during periods of revolutionary change. His view was that arguments are not by themselves logically compelling.
[26] Kuhn said that he modelled revolutionary change on his experience as a historian (RSS, 87). A historian, or individual scientist, can have these "aha experiences", but not a scientific community as a group. "Communities do not have experiences, much less gestalt switches" (RSS, 88).
[27] See SSR-4, 122n13 for the empirical evidence Kuhn cites and RSS, 57, for the distinction between having empirical evidence and making a logical point.

perfectly. It symbolized the logical point Kuhn wanted to make, that is, that successive paradigms are logically incongruous – what was anomalous before a revolution is normalized in a different logical structure afterwards, and one cannot move logically from the one to the other. What is more, by prioritizing the logical level, Kuhn distanced himself from an individualistic understanding of creativity that emphasizes psychological characteristics. Instead of talking of geniuses, of individual talent and individual experiences, he focused on communities, on institutional practice and logic.

Kuhn's most explicitly stated view on creativity appears in his paper "The Essential Tension: Tradition and Innovation in Scientific Research" (ET, 225–239). This paper is based on a talk Kuhn gave at a conference of psychologists on the identification of creative scientific talent. Psychologists and educators then, and now,[28] insist that creativity is enhanced by divergent thinking. Kuhn stressed in that article the significance of convergent thinking as well for scientific advancement. He said that it is vital, or essential as the article's title has it, for scientific research to support the tension between these two conflicting modes of thought which ought to characterize both the scientific community and the individual scientists (ET, 226, 227–228n2). "The productive scientist must be a traditionalist who enjoys playing intricate games by pre-established rules in order to be a successful innovator who discovers new rules and new pieces with which to play them" (ET, 237). Why is convergent thinking so important for innovation in science? Because "novel discoveries in the mature sciences are not born *de novo*," they emerge when a well-trodden field is transformed by new patterns of organization (ET, 234). This is similar to the way a new aspect emerges from the same data. Innovative thinking in science is made possible by the previously held beliefs within a matrix of expectations. Scientific education and institutional research practices define the area that scientists work in and prepare them to deal with the difficulties they may encounter. "In the mature sciences the prelude to much discovery and to all novel theory is not ignorance, but the recognition that something has gone wrong with existing knowledge and beliefs" (ET, 235). The same thought is expressed more rigorously in SSR:

> Novelty emerges only for the man who, knowing *with precision* what he should expect, is able to recognize that something has gone wrong. Anomaly

---

[28] For a review of the literature on divergent thinking, see Runco (2010). Cf. Sir Ken Robinson's talk (2010) on the significance of divergent thinking in education.

appears only against the background provided by the paradigm. The more precise and far reaching that paradigm is, the more sensitive an indicator it provides of anomaly and hence of an occasion for paradigm change. (SSR-4, 65)

Anomaly literally means deviation from normalcy.[29] Only those acquainted with what is normal, in all its rigor and detail, through education and professional practice, can recognize concrete trouble and offer remedies that are apt and relevant. As Kuhn says in "The Function of Dogma in Scientific Research" (1963, 349), commitment to received scientific tradition "provides the individual scientist with an immensely sensitive detector of the trouble spots from which significant innovations of fact and theory are almost inevitably educed." If scientists do not know the field and are willing to simply try new ideas, they will return their science "to its preconsensus or natural history phase" (ET, 234; cf.), where there is rampant disagreement over fundamentals and no development. Kuhn subscribes to the Baconian methodological dictum that "[t]ruth emerges more readily from error than from confusion" (SSR-4, 18).

Creativity has been standardly associated with freedom, imagination, spontaneity and natural talent – all of them attributes of individuals. But, in contrast, Kuhn insisted on the institutional framework that inculcated and promoted a culture of discipline and commitment. What others took to be inevitable human limitations of individual scientists, he considered institutional preconditions of success (Kuhn 1963, 348–349). Creativity for Kuhn required a balance between deep, even dogmatic, commitment to the status quo and the professional ideology of innovation and freedom of exploration (ibid., 368–369).

Acknowledgement

I would like to thank Brad Wray for his comments and suggestions which helped me improve the paper and Mattie Clear, from the MIT Libraries, Department of Distinctive Collections, for her assistance in locating archival material.

---

[29] Anomaly comes from the Greek word "anōmalia" which is the noun of "anōmalos", formed by the privative prefix *an* and *homalos* which means even, regular. So Hacking wrongly states, in his introduction to the fiftieth anniversary edition of SSR-4 (xxvi), that "anomaly" is formed by the privative prefix *a* plus *nom* which comes from the Greek word for "law" (he means the word "nomos").

CHAPTER 10

Kuhn on Scientific Discovery as Endogenous
Thomas Nickles

10.1 Introduction: Taming Discovery

One of Thomas Kuhn's major purposes in SSR, one often overlooked, was to provide an endogenous philosophical account of scientific discovery[1] – an impossible task according to the received view. No one claims that all discoveries are endogenous, but the received view excluded all of them as exogenous to "the logic of science." Far better known is Kuhn's attack on then-standard accounts of scientific progress as incremental and cumulative, gradually approaching a final truth.

"Endogenous" and "exogenous" are biological and economic terms. Earthquakes that wreck major industrial sites are exogenous bad luck to the economy, a bolt out of the blue. Today innovation is widely recognized as a main driver of economies, so it is surprising to learn that economists paid little attention to innovation until the last decades of the twentieth century. Before that, they commonly regarded an innovation as a development external to the normal work of a firm, an industry, or of an entire economic system, not as something that can be given an internal economic explanation.

I thank Brad Wray for the invitation to contribute. I have learned much from his work and from that of experts such as Paul Hoyningen-Huhne (1993), Alexander Bird (2000), Hanne Andersen (2001), Theodore Arabatzis (2005), Vasso Kindi (2012), Howard Margolis (1987), Ron Giere (1988), Joe Rouse (1996), Nancy Nersessian (2008), and Rogier de Langhe (e.g., 2017). Bill Wimsatt's work (e.g., Wimsatt 2007) has always stimulated my thinking.

[1] Kuhn used "discovery" and other terms to fit the language of the day. Although a discovery is an achievement within an expert community, he did not use "discovery" as an achievement term in the strong sense that a final truth about the universe had been established. He occasionally spoke of "invention" when a theoretical interpretation was attached to the raw discovery of a phenomenon or entity such as oxygen (SSR-2, 66, 79). I shall follow his metaphysically noncommittal use of "discovery" and shall sometimes substitute "innovation" for both "discovery" and the development of new equipment and techniques.

Meanwhile, in mid-century philosophy of science, it was also widely accepted that innovation at scientific frontiers is the font of new scientific knowledge. For example, Karl Popper asserted that the problem of the growth of knowledge is the central problem of philosophy of science. Yet, despite the title of his most famous book, *The Logic of Scientific Discovery* (1959), he joined the logical positivists and logical empiricists in denying that there can be a logical or rational or epistemologically relevant account of discovery. For these methodologists, there was no such thing as a "logic of discovery," only a "logic of justification." In Popper's case, not even that, since Popper denied that universal scientific claims can be justified inductively.

SSR was, among other things, Thomas Kuhn's attempt to endogenize scientific discovery within a philosophical account of scientific research. Yet Kuhn, too, rejected the existence of a logic of discovery and that of a master scientific method more generally. How did Kuhn do it? How did he accomplish the trick of being a friend of discovery while rejecting a logic of discovery and even "the scientific method"?

Kuhn did this in several interlocking steps.

(1) He rejected the view that scientific discovery is a sudden, unanalyzable "aha" experience of individuals. Instead, it is an articulated, analyzable process based on problem-solving strategies that typically involve many researchers, as we learn when we zoom in and do the history in careful detail (ET, chapter 7, 165–177). In effect, Kuhn's move reduces macro-discoveries to a series of mini-discoveries (see Section 10.4, on problem reduction). The discovery of oxygen is his best-known example (SSR-2, 53ff).[2]

(2) He rejected the existence of a quasi-algorithmic logic of discovery, that is, rules for accomplishing innovative research (SSR-2, chapter V).

(3) He regarded research as a problem-solving activity. "[T]he unit of scientific achievement is the solved problem" (SSR-2, 169). And he had much of cognitive-epistemological interest to say about how mature scientists articulate and address problems.

---

[2] Page and chapter numbers are to the second (1970) edition of SSR, which includes "Postscript-1969." Later editions are basically reprints. Often the person perceived to be making the original breakthrough is retrospectively credited with a discovery that took years to articulate. Kuhn's oxygen example hints at this. Another example is the quarter century of new and ever-more-illuminating derivations of Planck's black-body radiation formula of 1900. Indeed, the burden of Kuhn's quantum history (BBT) is that "Planck's" discovery was mostly the work of Einstein and others in 1905 and later. Good, anti-whig historical work has disclosed that, ironically, many discoveries were completed long after "their" original discovery.

(4) This is especially true of the restricted, well-structured "puzzles" that characterize "normal science" in the exact sciences.
(5) The overarching "paradigm" practically guarantees that the key problem-solutions and devices that Kuhn termed "exemplars" provide sufficient resources to solve legitimate research puzzles via an intuitive, pattern-matching process primed by "acquired similarity relations" (SSR-2, chapters IV, V, Postscript).
(6) In contrast to traditional logicians of science, who shunned rhetoric as imprecise and subjective, Kuhn introduced rhetorical relations (analogy, simile, metaphor) as key to the expert perception of pattern matches and mismatches.
(7) Kuhn's major epistemological innovation was to replace rule-based method (hence, rule-based learning) by a special sort of learning-by-example acquired by problem-solving experience as a student and then as a researcher. These experiences instill a technical, domain-specific intuitive capacity that enables the rhetorical matching process (SSR-2, chapter V, Postscript).
(8) Surprisingly, Kuhn claimed that genuinely exogenous discoveries are unwanted in normal science. Being unexpected, they are disruptive and hence discouraged. They challenge the overarching macro-paradigm[3] and are thus, potentially revolutionary. Excluding them, or at least taming them, becomes an urgent matter (SSR-2, 35). (There is an irony here. Taken in fully tamed seriousness, Kuhn's normal-scientific account of discovery leaves no place for major discovery! Only already expected results are welcomed.)
(9) Moving beyond the received debate over "logic of discovery" and "logic of justification," Kuhn emphasized the generative and justificatory roles of heuristic fertility, of heuristic attraction, in his more forward-looking account of research. Researchers' assessments of heuristic fertility are crucial to Kuhn's account of both normal science and the transition to a new paradigm. The point of normal science, he says, is to cash in its heuristic promise (SSR-2, 24). And it is the heuristic promise of exciting new work that attracts people to an emerging new paradigm when normal science becomes either sterile or problematic. In a crisis, heuristic promise signals a possible route toward saving the field.[4]

---

[3] I shall henceforth use the terms "macro-" and "micro-paradigm" to capture Kuhn's distinction, in his "Postscript-1969," between disciplinary matrices and exemplars.
[4] Kuhn presents crisis as a time of great worry and anxiety among leading researchers, quoting Pauli (SSR-2, 83f). But, for many, it is a time of great excitement, precisely because it suggests that major new work can be done. Today, for example, there are discrepancies in measuring the Hubble

In sum, Kuhn tamed the problem of understanding discovery in his account of normal science in order to treat it as a rationally articulated process. But he then had to compensate by positing a correspondingly wild, radical view of paradigm change, and this move exacerbated the discovery problem at that locus. For he then had to admit that he had no idea where the uprooting new ideas and techniques come from (ET, 332f). Thus, he failed fully to endogenize discovery.

While Kuhn's account of mature science is full of illuminating insights, there are also exasperating ambiguities and tensions. For example, under critical fire, he took "exemplar" to be the primary sense of "paradigm" (SSR-2, Postscript). Although rejecting the spontaneous "aha!" psychology of discovery, he retained gestalt switches, to some degree, in the normal-science recognition of a pattern match and, more obviously, in his account of revolutionary change, but he soon qualified the latter view as well.[5]

Some of these tensions can be resolved by realizing that Kuhn was working on three levels with three distinct points of view: anti-whiggish historian of science, philosopher of knowledge, and working scientist. He distinguished his historical work from his philosophical work (ET, chapter 1, 3–20). In addition, SSR contains two quite different points of view: Those of the researchers working in the scientific trenches, convinced that they are on the track of truth about the universe, and Kuhn's own view as a historical philosopher, standing above the fray and regarding their work as progressive, but as leading inevitably to the next scientific revolution.

10.2 Normal Science

How, more exactly, does normal science tame the discovery problem? Kuhn's basic answer is that it trades in relatively small, well-defined

---

constant by different but extremely precise methods, with standard cosmological theory at stake. Here are some typical comments (from Achenbach 2019).

> No one is panicking. To the contrary, the theorists are intrigued. They hope the Hubble Constant confusion is the harbinger of a potential major discovery – some "new physics." "Any time there's a discrepancy, some kind of anomaly, we all get very excited," said Katherine Mack, a physicist at North Carolina State University who co-wrote a recent paper examining the issue.

Insofar as such excitement over challenges to the current orthodoxy is widespread, Kuhn's psychological profile of normal science practitioners is badly mistaken.

[5] Kuhn soon realized his confusion of individual gestalt switches with (impossible) community gestalt switches. He was attracted to Gestalt theory as one of the few cognitively interesting perceptual psychologies then available. But he then pretty much dropped psychology, and also sociology, from his internalist, intellectualist account. See Galison's (1981) excellent review of Kuhn (BBT). For suggestions about lines Kuhn could have taken, see Giere (1988), Nersessian (2008), and Nickles (2000 and 2003), as well as the items cited in the acknowledgements.

problem-solving episodes. For that reason, Kuhn terms the targets of normal research "puzzles" rather than problems (SSR-2, chapter IV). The term derives from a comparison with puzzles such as jigsaw puzzles. The latter present us with highly constrained problems, the solvability of which, within the established practices, is guaranteed in advance. Similarly, for Kuhn, normal science under a macro-paradigm is such a tradition-bound enterprise that it guarantees the solvability of the puzzles that arise within it, in fact, solvability in terms of the resources that the specific research tradition has already provided (SSR-2, 37). The primary resources are the exemplars, which include key tools, such as measuring instruments (SSR-2, 187) as well as the kinds of problems-cum-solutions used to educate science students and that appear in key technical papers. Given Kuhn's emphasis on the guarantee of solvability under the paradigm, I have previously used the algebraic metaphor that the set of exemplars is sufficient to span the problem space (or puzzle space) of the paradigm.[6] Or, to put the point in the post-Kuhnian language of Donald Norman's affordances, the exemplars provide affordances, hooks, handles, points of view, attractors that invite researchers to approach specific types of puzzles in specific ways.[7] Intuitions acquired as a student and tuned by professional experience provide the research expertise that directs attention toward promising lines of work.[8]

In Kuhnian normal science, the problems and solutions are small enough that, given the constraints on them and the expertise of the investigators, discovery no longer seems like a gift from the gods or the powers of genius. Yet normal science only takes discovery this far. There is nothing like an algorithmic logic of discovery available here. It can still require great talent, and sometimes luck, to solve normal scientific

---

[6] See Nickles (2003 and 2012) for a fuller account of normal science as well as Kuhn's somewhat problematic concept of exemplars. On exemplars, see Shan (2018). See also William Goodwin's chapter in the present volume for his account of Kuhnian normal science.

[7] As we once sat together in a cognitive science lecture, Kuhn agreed that "point of view" could be important in solving a problem such as the missionaries and cannibals problem. On affordances, see Norman (1990, 1993). The term "affordance" is ambiguous in the literature, depending on whether an affordance is regarded as a feature intrinsic to an item (as it was for perceptual psychologist J. J. Gibson) or is a matter of the relation of the item to an agent with specific physical and cognitive capacities (as it was for Norman). Although she does not use the term, Kelly Hamilton's (2001) fascinating article on Wittgenstein's engineering education indicates how his schooling tuned or primed him to view problems in specific ways, e.g., as solvable by particular combinations of gears.

[8] Despite becoming a major (but reluctant) influence on the newly emerging sociology of science, Kuhn remained a dedicated internalist for the rest of his life (Kuhn 2000b). Today it is clear that contextual constraints help shape scientific research. These include national priorities, funding sources, journal orientations and rankings, local lab priorities and cultures, and much else. These, too, have a heuristic, directive function, in a broad sense of "heuristic."

puzzles.[9] There do, after all, remain creative moments that may involve an individual's "aha" experience of a pattern match via learned intuition. Kuhn thought he was better than anyone in his generation at "getting inside the heads" of key practitioners (Kuhn 2000b, 276), to follow their lines of thought and action. But when it comes down to it, no matter how much we say about a given person's cognitive preparation and her socialized and routinized practices, we cannot get beyond high plausibility in explaining why a particular idea or action occurred to her. That fact surely holds for most explanation in human affairs.

Two long-term influences of Kuhn's account of exemplars in research are worth noting here. First, exemplars are models. Along with that of Mary Hesse (1961, 1963/1966) and a few others, Kuhn's work invited philosophers' increased attention to models and modeling at all levels of research with correspondingly less emphasis on high theory. Second, in focusing on research processes, where his predecessors were mainly interested in finished products, Kuhn's work helped to activate the "practice" movement in science studies, including, eventually, philosophy of science.[10]

10.3 Where Do the Basic Exemplars Come From?

Although he drops some hints, Kuhn never really answers this question endogenously in SSR and related essays, as far as I can see. If so, we have a finer-grained explanation of why his account of discovery is not entirely endogenous. On a straightforward reading of his work, the basic exemplars cannot be endogenous to normal science, since a given regime of normal science already depends on their existence. This leaves only the crisis period that may lead to revolution, a time when normal science is breaking down and deviations from normal science are increasingly permitted to solve resistant problems. Unfortunately, the iconoclastic departures from the old paradigm are precisely where Kuhn throws up his hands in defeat. For the most basic exemplars that eventually crystallize a new macro-paradigm will

---

[9] This limitation on "discovery" is not a defense of the old views about logic of justification, for, as an ongoing practice, finding a suitable mode of confirmation (including constructing the equipment and the arguments) is itself a problem-solving, "discovery" procedure that often requires great talent. The same is true of constructing any nontrivial logical argument. The premises and steps do not simply present and then organize themselves. Most historical explanation is arguably no better than this.

[10] However, despite the recent, healthy emphasis on scientific practice, "practice" remains a vague term, given that so many variables are in play, many of them hidden from us (Daston 2016, 126; Wilson 2004).

be highly innovative, previously unexpected departures from the old macro-paradigm.

Exemplars are surely among the most important scientific discoveries or innovations for Kuhn, because the set of basic exemplars structures the domain of the macro-paradigm. (I am here speaking of the key, "founding" exemplars – my term – not the variety of specialized exemplars derivative from them that one expects to find in the practice of normal science, as it articulates the macro-paradigm.) The need for a new, basic exemplar within an ongoing normal science would conflict with Kuhn's claim that a macro-paradigm guarantees the solubility of any puzzle formulable in its terms. Indeed, the addition of new, basic exemplars would seem to change the macro-paradigm – although Kuhn can reply that, by "paradigm change" he means only the change to an incompatible paradigm, and that a paradigm remains the same paradigm as long as the new exemplars are compatible with the old ones.

Kuhn sometimes writes as if exemplars remain static in nature during the course of normal science, yet he surely would have to agree that they, too, can evolve as they are repeatedly articulated in new applications. Experts grappling with a research puzzle attempt to match it, at least partially, with one or more exemplars. Achieving a match may require deforming (reframing) somewhat both the existing exemplar(s) and the puzzle to be solved in terms of it or them. Success here is surely interactional: An exemplar basic to the solution may now have a deeper meaning via the new application, one that perhaps links two previously distinct ideas or practices.

The difficulty that I see here for Kuhn is this. Where do we draw the line between normal maturation or evolution of an exemplar and revolutionary replacement? Here the debate, traditionally over macro-paradigm-change, arises at the level of exemplars.

There is a related issue. Within normal science, Kuhn tells us, all experts solve problems by employing the same exemplars in the same ways. Thus, they can readily agree on a new puzzle solution, even when they differ in their interpretation of what the exemplars mean at a deep, realist level, for example, whether Newton's second law is a deep empirical claim or merely a definition (SSR-2, 44, 50, 180, 188; 2000b, 298). It is their practice that counts, not their private, metaphysical views. (Today's quantum theorists disagree wildly about the metaphysical reality underlying the theory, or even whether there is one, while working smoothly together on routine science.) But here, again, where do you draw the line? Often this difference comes to nothing in practice, but sometimes the underlying disagreement is a difference that makes a difference.

Why does such a difference not matter in normal science yet may make a revolutionary difference in the transition to a new paradigm? As noted, Kuhn's answer is that the professional constraints of normal science begin to break down in a crisis, permitting various degrees of scientific deviation. But is this not precisely one way in which a new, paradigm-changing exemplar can emerge from normal scientific work? In his fascinating lecture, "What are scientific revolutions?" (Kuhn 1987/2000), it would seem that revolutionary change (as Kuhn construes it) has arisen directly out of such differences. There he calls our attention to the consequential reinterpretation of Volta's electric pile and of Planck's resonators.

We can again speak here of the different perspectives as providing different affordances, changing the heuristic profile of future research options. To use the term that I apparently borrowed from Ernan McMullin (1976), it can make a difference in the heuristic appraisal creative scientists give to available research options. For example, young Einstein and Ehrenfest saw the heuristic potential in the idea of modeling a radiation cavity in terms of oscillators rather than Planck's resonators. This helped them to misinterpret Planck as addressing the blowup that Ehrenfest dubbed "the ultraviolet catastrophe," whereas Planck saw his work as an extension of normal, classical science (BBT; Klein 1970).[11]

These difficulties left Kuhn struggling to explain where the unorthodox new ideas and techniques come from, but what he said about acceptance is on the right track, I think. A radically new way of solving an old problem (no longer a mere puzzle, given the crisis) gets scientists' attention for three reasons. First, it provides a "how possibly" explanation or "proof of concept" of an extremely puzzling phenomenon or conceptual difficulty, provided that one is willing to suspend religious allegiance to part of normal science. This first pass can then be reverse-engineered, so to speak, in order to determine precisely why it worked, for example, to solve a first-order problem of explaining spectral lines or to avoid the ultraviolet catastrophe. In the process, generation and justification are closely intertwined. Second, it thereby provides new work for the experts in that domain, a future-oriented point that philosophers interested in

---

[11] The early quantum theory will be my running example, both because of Kuhn's book, BBT, and because I have dealt with the work of Planck, Einstein, Ehrenfest, and others in Nickles (1976, 1980, 2005) in a case study of the transition to the old quantum theory. Among secondary sources, I am much indebted to the papers of Martin Klein and to his book on Ehrenfest (Klein 1970), as well as to Kuhn's BBT.

confirmation theory often overlook. Third, this work offers a concrete hope of saving the field from its present impasse.

While the first reason challenges scientists' self-understanding, the second and third promise to restore their integrity and that of the field. After all, they can be creative scientists in the domain in question only insofar as they are actively doing creative work in that domain.

I will briefly mention two other worries about Kuhn's treatment of exemplars. First, sometimes his philosophy gets in the way of his history. For example, he assimilates the emergence of new sciences, along with their exemplars, to the classical, "Newtonian mechanics," even though some of these new exemplars (e.g., noncentral forces of electromagnetic theory and entropy fluctuations in statistical mechanics) were obviously incompatible with the classical "Newtonian" paradigm. So why don't these count as paradigm change? Sometimes Kuhn seems to allow that they do. At other times he takes the rather whiggish view that "classical physics," the modern physics up to relativity and quantum mechanics, was a single paradigm.

My second worry is that Kuhn's exemplars are all positive models for emulation. In practice, mature scientific fields also have negative exemplars – classic models of mistakes of various kinds, to be avoided. In his late work, Kuhn hinted that there could be negative exemplars (Kuhn 1987/2000).

10.4 Problem Reduction and Modeling

Kuhn attempted to kill the standard account of theory reduction then dominant, while championing what I shall call problem reduction. He denied that theory reduction occurs across revolutionary disruptions within a domain, whether the reduction is Nagelian or limiting. In Ernest Nagel's (1961) account, reduction is essentially the Hempelian explanation of one theory as a deductive, special case of another via the use of "correspondence rules" to link the premises (the language of the reducing theory) to the conclusion (that of the reduced theory). Nagel's leading example was the reduction of classical thermodynamics to statistical mechanics, where thermodynamic temperature corresponds to the mean kinetic energy of the molecules.

A more flexible account of reduction allows limiting and other, more complex relationships between the two theories. In this case we say that the later, more mature theory reduces to the earlier one rather than the earlier one being reduced by the later (Nickles 1973). One commonly made but oversimplified claim is that there are reductive relations of this sort between quantum mechanics and classical mechanics, as Planck's constant goes to zero. A cleaner example is the relation of special relativity theory to classical mechanics, as velocity goes to zero or the speed of light to infinity. Here Kuhn (SSR-2, 101) entered his major objection:

> There is a revealing logical lacuna in the positivist's argument, one that will reintroduce us immediately to the nature of revolutionary change. Can Newtonian dynamics really be derived from relativistic dynamics? What would such a derivation look like? ... [Here Kuhn sketches such a derivation in which velocities are restricted to low values, yielding conclusions N_1, N_2, etc., that look like Newton's laws of motion.]
>
> Yet the derivation is spurious ... Though the N_i's are a special case of the laws of relativistic mechanics, they are not Newton's Laws. Or at least they are not unless those laws are reinterpreted in a way that would have been impossible until after Einstein's work.

This is Kuhn's well-known meaning-change objection to then-standard accounts of reduction. Many critics have noted his controversial use of a holistic account of meaning here, tied to his gestalt-switch psychoanalytic internalism (Galison 1981). The point I want to make is different. Notice that in attempting to refute these accounts of theory reduction in order to defend his account of disruptive revolution, by turning the logical tools of the positivists against them, Kuhn himself employs those very tools and thereby ignores the forward-looking heuristic power of the rhetorical tools that he touts elsewhere in SSR. Despite his overall emphasis on the importance of heuristic fertility and the resemblance intuition, Kuhn here failed to appreciate the power of limiting and other relationships between theories.[12] He was too focused on using their own tools to undermine the logicians' arguments.

Pressing this point further, we may regard problem reduction as one key to the success of Kuhn's account of discovery in normal science. Problem reduction is the main route to routinized problem-solving; hence, it amounts to a sort of method without rules and without needing derivation from high theory (SSR-2, Postscript). Under relevant rhetorical operations such as analogy, a research problem is found to reduce to an established exemplar or a combination of exemplars. So another way to formulate my complaint is this: Kuhn rejects theory reduction across revolutions on strict logical grounds, thereby defending his account of revolution, while defending normal-scientific problem reduction on far more flexible grounds.[13] I see a bit of irony here!

---

[12] See Nickles (1973, 1976, and 2005).

[13] I tried to provide an early account along these lines in Nickles (1973, 1976, 2005), where I distinguished two kinds of intertheoretic reduction, domain-combining reduction (reduction$_1$), which results in ontological unification of previously distinguished entities or processes (e.g., of light

There are many kinds of problem reduction. Anything that reduces the size of the problem space or search space for investigators counts as problem reduction in this broad sense. Indeed, the transformation of a problem into a logically equivalent problem that is more tractable, either intuitively or computationally, still counts as a reduction of the search space relative to the cognitive capacity of the researchers.[14] The point of much modeling is problem reduction, which is why Kuhn welcomed the rhetorically loaded work on models by N. R. Hanson (1958) and Mary Hesse (1961, 1963/66). One sort of problem reduction is to reduce the number of independent variables. In various sciences, one faces problems of high dimensionality and employs various techniques to reduce the number of variables in play. This can be done nonrigorously, as in approximate models and simulations, or rigorously, as when Gustav Kirchhoff showed how to reduce the black-body problem of finding an equation relating the distribution of energy intensity over frequency v and temperature T to the problem of specifying a function of a single variable v/T. In the black-body case, others such as Stefan, Wien, and several experimentalists later provided further constraints that imposed limit relationships or bounds, each of which further defined and thereby reduced the problem (Nickles 1980, 1981; Galison 1981).

At a higher level of description, we can say that Kuhnian normal science was Kuhn's attempt to reduce the problem of explaining scientific discovery to that of how puzzles are solved, that is, reducing problems to highly constrained puzzles.[15] Kuhn's major epistemological and methodological innovation was surely his account of direct rhetorical modeling of puzzles on those problems solved by exemplars. Again, when faced with a new puzzle, the first stage is to find resonances to already available exemplars and then to articulate these into a defensible solution to the puzzle. These moves reduce the puzzle to seeing it, perhaps partially, as a member of one or more classes of already solved problems.

---

with electromagnetic waves, heat of a gas with mean kinetic energy of the molecules) and domain-preserving reduction (reduction$_2$), in which a successor theory replaces a predecessor, without appreciable cross-domain reduction. For example, we have reduction of special relativity to the earlier, classical mechanics in the limit of low velocities versus reduction of the earlier, physical optics by the seeming domain-distinct electromagnetic theory, although there exist some limit and other relations here, too. Reduction$_1$ is "horizontal," while reduction$_2$ is "vertical."

[14] On reductive heuristics, see Wimsatt (2007), chapter 5 *et passim*. Kuhn's exemplars play a heuristic role in this broad sense. Reductive techniques can be material as well as linguistic/symbolic, as in organic chemistry. See, e.g., William Goodwin's chapter, this volume.

[15] During the same years, Herbert Simon was also endogenizing scientific discovery by focusing on problem-solving, about which experts already had much to say. Simon took a computer-science and cognitive psychology approach.

Kuhn's most extensive example brings out the rhetorical, "similarity relations" component of problem reduction explicitly, a process that involves "minimal recourse to symbolic generations" from high theory. Galileo came to see balls rolling down and then up an inclined plane as "like" the motion of a simple pendulum, to use Kuhn's expression. This became the key exemplar. Omitting details, Huyghens was then able to solve the problem of the center of oscillation of a physical pendulum by modeling the physical pendulum as a composition of Galilean point-pendula. This was a reduction operation. "Finally, Daniel Bernoulli discovered how to make the flow of water from an orifice resemble Huyghens' pendulum. . . . From that view of the problem the long-sought speed of efflux followed at once" (SSR-2, 190).

Ian Hacking (1983) and other writers on experiment have helped us to recognize that there are experimental traditions independent of then-current high theory. As I read him, Kuhn had earlier made a similar point in insisting that problem-solving can occur (especially in its reduced form as puzzle-solving) with minimal recourse to derivations from higher theory and in recognizing that there are lineages of problems-cum-solutions, as in the pendulum-efflux problem just described. That insight is at one with his claim that scientists learn by example rather than by rule or by top-down theoretical derivation. This means that problem reduction can proceed independently of large theory reductions. In a case study of the transition to the early quantum theory, I argued in detail that that is how Ehrenfest proceeded when trying to isolate and extend the key features of the new problem solutions.[16] In a sense, he methodized the extension of specific exemplars. In particular, his adiabatic principle reduced the problem of finding the quantum conditions for whole classes of mutually adiabatically transformable physical systems by showing that determining the quantum conditions for any one of them was sufficient to cover them all, independently of theory reduction.[17]

On that note let us turn to Kuhn's historical treatment of precisely that transition. Is his history of the early quantum theory, BBT, compatible with SSR?

---

[16] For details, see Nickles (1976, 2006). Post (1971), building on the "conservative induction" tradition, speaks of building new theories by finding the "footprint" of a new theory in the flaws of the old and by exploiting invariances and correspondences.

[17] Compare NP complete problems in computational complexity theory. Each problem can be reduced to each of the others in polynomial time. If there is an algorithm that can solve an NP complete problem in polynomial time, all NP complete problems can be solved in this way, but currently there is no known way to solve any of the problems quickly, that is, in polynomial time.

10.5 Continuity or Discontinuity? The Early Quantum Theory

We have seen that, although Kuhn did say many suggestive things toward endogenizing scientific discovery, his treatment of scientific revolution set up as many barriers to such an account as his treatment of normal science removed – and for that very reason. As a result, his endogenous treatment of discovery remained pretty much stuck in the cumulative conception of science that he wished to attack. For he agreed that progress in normal science is cumulative. I shall end on a positive note by joining those who point out that, had Kuhn softened his sharp distinction of normal from revolutionary science, he would have had resources for addressing the bolder discoveries. In a sense, he did already possess those resources, as his Postscript to SSR and his excellent history of the old quantum theory, BBT, reveal. Of course, the resulting revision of SSR would have blurred his normal/revolutionary distinction and thereby made his model of science less arresting.

The revolutionary discontinuity-versus-continuity debate has led to an impasse, with good points made on both sides. Two of the major issues concerning discontinuity in Kuhn's work are whether he is correct that major scientific developments, especially rapid ones, are sufficiently discontinuous with preceding work to be called revolutions, especially "revolution" in the sense of a dramatic overthrow and replacement of the previous framework – one that is so great that the two sides can no longer communicate effectively.[18] The other issue involves the particular case of the transition from classical mechanics to quantum theory. Critics (especially Klein et al. 1979) have claimed that Kuhn's detailed history of the early quantum theory – *Black-Body Radiation and the Quantum Discontinuity 1894–1912* (BBT) – is incompatible with the discontinuous account of normal science and revolutions found in SSR. Interestingly, Kuhn neither cites SSR nor employs his normal-revolutionary distinction in BBT. As noted, he always attempted to keep his historical work insulated from his philosophical work.

In this transition, do we have continuity or discontinuity? Surely the answer is yes and yes! Kuhn himself may have come to a version of this conclusion. We may regard the aforementioned, later lecture, "What Are Scientific Revolutions?" (Kuhn 1987/2000) as a bridge between SSR and

---

[18] On incommensurability, see Nickles (2017) and the chapter in this volume by William Devlin.

BBT, where he does cite the latter for a fuller treatment of his crucial example of Planck's work.[19] Three of the indicators of revolution that he develops in that paper are (1) its holistic nature (versus point by point, cumulative change), (2) meaning change ("change in the way words and phrases attach to nature") resulting in taxonomic change (placing items in a different category than before, e.g., what counts as a planet, star, and satellite), and (3) a crucial change in the relevant rhetorical relations – what is similar or analogous to what? These were topics on which Kuhn worked to the end of his life, especially with his biological-speciation-inspired account of specialization (see, e.g., Wray 2011 and both Wray and Barker in this volume).

While Kuhn insightfully recognized that there are linguistic and other cultural differences between fields, even between nearby specialties, it is also true that scientists often find ways to bridge those differences (see, e.g., Galison 1997, chapter 9, on trading zones and pidgin languages). Creative scientists working at frontiers are prepared, both technically and emotionally, to navigate new and unfamiliar terrain. They often possess the mentality, the self-confidence, of bold explorers. If early Kuhn were absolutely correct about incommensurability, today's large research teams with people from many disciplines, none of whom has complete understanding of the experiment or associated theory, would be quite impossible. Kuhn may have realized this when he noted that the SSR story no longer applied well to the mélange of new, interlocking specialties and the Big Science that emerged during World War II and after. This was a strange admission about the end of a historical era, given the somewhat a priori character of Kuhn's picture of scientific development as alternating phases of tradition and innovation (Kuhn's "essential tension"). The early Kuhn already spoke of "the necessity" of scientific revolutions (SSR-2, chapter IX) and often described himself as "a Kantian with moveable categories" (see Lydia Patton's chapter in the present volume).

We can put the resources point in terms of exemplars. Kuhn provided an important clue in his "Postscript" account of the early seventeenth-century solution of the pendulum-efflux problem, mentioned in Section 10.4. We can speak of a genealogy or lineage of exemplars, ranging across different branches of science, in this case from Galileo's work on inclined planes through Huyhens on physical pendula to Bernoulli on the flow of fluid from the orifice of a tank. Given that Kuhn's direct modeling relation is a rhetorical relation rather than a logical relation, we can see many such

---

[19] Kuhn himself treats it as a bridge in his work (2000b, 314).

lineages linking classical mechanics and mature quantum theory via the old quantum theory, for example, successively deeper reinterpretations of Planck's formula and of quantum conditions, and of quanta themselves, of work on spectral lines, specific heats, and much else.

Recall that one of Kuhn's arguments for discontinuity is the meaning-change route to incommensurability. His primary example is the failure of special relativity theory to reduce to classical mechanics in the limit of low velocities (SSR-2, 101f). If we accept his early, controversial, holistic conception of meaning (e.g., the meaning of "mass" is implicitly defined by the equations of the theory), then Kuhn is correct that the deductive argument for theory reduction fails. But that is surely beside the point when it comes to genealogies of exemplars, for here rhetorical relations such as analogy and similarity take precedence over strict logic. It is surely not a stretch to see that particular classical motion equations have specific relativistic counterparts that play a similar role. From this point of view, the discontinuity becomes manageable. In the case of relativity, even the conservative Planck could quickly embrace it. And although Planck rejected the existence of free quanta for years after Einstein introduced the idea in 1905, it was because he thought the idea unnecessarily radical, not because he failed to understand it.

We should avoid monolithic, big-picture accounts with all-or-nothing outcomes such as requiring a totally new and unexpected set of exemplars.[20] It is not difficult to extend Kuhn's work on exemplars in normal science to achieve the sort of research continuity that he reports in BBT while recognizing that the eventual result is a body of work that looks strikingly different from classical theory. The very Kuhnian point that even radical scientific innovation is articulated is one key. We can view the emergence of mature quantum theory as a Big Discovery or Invention consisting of a long series of quasi-continuous steps (smaller innovations), the end points of the series being a world apart. Rapid evolution of this sort becomes indistinguishable from revolution.

Recent work by authors such as Robert Batterman (2002), Alisa Bokulich (2008), and Eric Winsberg (2010) go further than accounting for the possibility of a smooth, one-way passage from old theory or model to new. Starting from the post-revolution end, we see that Kuhn's account of revolution as complete overturning and replacement is too radical. In fact, in the quantum case, classical theory still finds essential use in several

---

[20] Kuhn's theory-centered account of monolithic paradigms retains much of the Neo-Kantian talk of "conceptual schemes" of the 1960s.

quantum problem contexts, and not only as a more tractable calculational device. Batterman focuses on universality in physics, while Bokulich provides examples from so-called quantum chaos, stressing how classical physics is essential to getting the quantum treatment right in the "mesoscopic" regime of phenomena.

> Semiclassical mechanics can be broadly understood as the theoretical and experimental study of the interconnections between classical and quantum mechanics. . . .
> There are three primary motivations for semiclassical mechanics: First, in many systems of physical interest, a full quantum calculation is cumbersome or even unfeasible. Second, even when a full quantum calculation is within reach, semiclassical methods can often provide intuitive physical insight into a problem, when the quantum solutions are opaque. And, third, semiclassical investigations can lead to the discovery of new physical phenomena that have been overlooked by fully quantum-mechanical approaches. Semiclassical methods are ideally suited for studying physics in the so-called mesoscopic regime, which can roughly be understood as the domain between the classically described macro-world and the quantum mechanically described micro-world. An area in which semiclassical studies have proven to be particularly fruitful is in the subfield of [so-called] quantum chaos. (2008, 104f)

Not only is the old theory not eliminated, but also it crucially helped the development of the new and is in turn illuminated by the new work. Moreover, it continues to help make the new theory intelligible and thus continues to act as a heuristic guide. This seems directly contrary to Kuhn on incommensurability; for here, at the site of one of the most dramatic revolutions in the history of science, the old theory helps us make sense of the new! Kuhn's early, holistic, gestalt-switch conception of an individual's appreciation of revolution ("the scales fell from his eyes") too quickly assumed that the new theory was highly intelligible in a self-contained manner, in isolation from its past, once one got inside it.

10.6 Concluding Summary

Kuhn's account of scientific discovery in SSR and related writings was incomplete and raises unresolved issues, but it helped to endogenize, and thereby to legitimize, scientific discovery as a topic for philosophy of science. It is by now clear that there is a great deal that we can say about scientific innovation and about things we can do to foster it, things more fruitful than dreaming before the fireplace, having black coffee on the balcony, or stepping onto buses.[21]

Kuhn's major epistemological innovation was his account of learning by example via a highly tuned rhetorical component of cognition, acquired in student problem-solving and later research practice, rather than by symbolic rules or top-down logical derivation from high theory. This was a general theory of learning, quite distinct from other accounts in the field, not only an account of scientific innovation.[22] While his account of scientific discovery was only partly successful, it focused attention on the then-badly neglected topic of scientific work as an ongoing process rather than a finished product. This emphasis, in turn, helped to stimulate "the practice turn" in philosophy of science that has taken root in recent decades.

---

[21] I am alluding to accounts of discovery by Kekulé, Poincaré, and others. See Hadamard (1945).

[22] I cannot here enter into the dispute about how much Kuhn borrowed from Polanyi (1958, 1966). While there are some strong similarities, there are also significant differences. And, in the 1960s, Skinnerian reinforcement learning showed, in principle, how such behavioral abilities could be formed without rules, based solely on experience with reinforcement. However, Kuhn was attracted to the new cognitive psychology of the time and to psychoanalysis, not at all to behaviorism. Unfortunately, he showed little interest in later developments in cognitive science and computer science, ones such as schema theory, expert systems, cased-based and model-based reasoning, connectionism, and genetic algorithms (Nickles 2000, 2003). The question remains open today whether, and to what extent, subconscious psycho-social processes can illuminate scientific discovery.

CHAPTER 11

Truth, Incoherence, and the Evolution of Science
Jouni-Matti Kuukkanen

11.1 The Concept of Evolution in Kuhn

The concept of the evolutionary development of science is one of the most important notions in Kuhn's philosophy. He emphasized its significance throughout his career but felt in the end that it "should have been taken more seriously than it was; and that *nobody* took it seriously" (RSS, 307).[1] Kuhn is correct in that scholars neglected the analysis of this concept for a long time in history and philosophy of science and, in particular, have not sufficiently attempted to see its role in the framework of his philosophy, while an unknown quantity of ink has been spilt over the earlier Kuhn's idea of radical scientific revolutions.[2]

I concentrate on spelling out Kuhn's rationale for pursuing the evolutionary approach. My view is that Kuhn first concluded that the historical record does not support the assumption that science develops teleologically towards the truth. This led him to outline an evolutionary view of scientific development. Kuhn argues that the future of the sciences is open-ended and that the sciences are bound to diversify, not unify; to become more fragmented, and not more integrated. To slightly adapt Kuhn's own terminology, the negative incentive that "pushed" Kuhn to outline an evolutionary image of scientific development is the lack of historical

---

[1] Cf. Kuhn's words about the significance of the evolutionary development in different stages of his career, e.g., SSR-2, 171; RSS, 307; and in his unpublished manuscript PW, chapter 1: "Scientific Knowledge as Historical Product", 13. The manuscript is still in the archives at the time of writing.
[2] Fortunately, the situation has already changed somewhat in recent years, as publications studying the evolutionary concept in Kuhn have emerged. See Wray's book *Kuhn's Evolutionary Social Epistemology* (2011), in particular part II "Kuhn's Evolutionary Epistemology" (79–143); Gattei (2008, 160–163, 168–172); Marcum (2015b, in particular chapter 6); Politi (2018); see also Kuukkanen (2012).

Truth, Incoherence, and the Evolution of Science 203

evidence for the teleological view of scientific progress. This is dubbed as the "negative rationale" in this essay. The positive elements, called the "positive rationale", that "pulled" Kuhn are the possibilities that the evolutionary image offers for explaining the dynamics of science.

While Kuhn emphasized the problems of the teleological view in his early writings, and in particular in the postscript of SSR, both the negative and positive rationales are visible throughout Kuhn's *oeuvre*. However, there is precious little writing about the evolutionary process of scientific development that would be philosophically illuminating in the middle of Kuhn's career. Only in some of the later essays does he devote more space for spelling out his conception of the evolutionary development.[3] It is a great pity that Kuhn did not have time to finish his book *Plurality of Worlds* (PW), which bore a promising working (sub)title: *An Evolutionary Theory of Scientific Development*.

In this chapter, I focus first on the reasons that make the teleological view unwarranted (Section 11.2). I then explain how the evolutionary alternative remedies the failures of the teleological model (Section 11.3). This discussion is continued in the subsequent section (11.4) of this essay examining the role of "niche" in Kuhn's philosophy.

The third main issue of this essay deals with the relation of Kuhn's evolutionary model to two concepts: truth and (in)coherence. I will examine what exactly Kuhn denied when he exclaimed that science is not converging on the truth and consider its relation to a central developmental notion: "niche" (Section 11.5). It turns out that the unattainable "truth" in the history of science signifies a comprehensive and correct, true, account of the world. If the historical process of the sciences leads to an outcome that consists of a large number of unintegrated disciplines, then there is no coherent world view in the end of this process, which could be true of the world in toto. Here, one can think of a layman's conception of an evolutionary tree (RSS, 99) where the trunk and the branches correspond to the developmental lines of the sciences and the leaves in the branches to all different kinds of scientific disciplines as the outcomes of these developmental processes. Of course, the leaves of this historical tree of the sciences may be as radically different as quantum physics is to art history. The entailment of the evolutionary view is that different branches have relatively few and even no links between them in the cases of distant disciplines, except perhaps historical links. To express this in more abstract terms, the

---

[3] Specifically, the essays "The Road since *Structure*" and "The Trouble with the Historical Philosophy of Science". Both are published in RSS in pages 90–105 and 105–121, respectively.

developmental lines form an incoherent set of approaches, theories and beliefs about the world with no unity in sight.

Kuhn not only thought that the sciences fail to arrive at the final Truth about the world with capital "T", but that truth as correspondence to the mind-independent world must be rejected. Because the niches, in which sciences are practised, change, there is simply no fixed and permanent mind-independent world to which statements and theories could correspond. However, it is unclear with what notion of truth Kuhn wished to replace the correspondence theory. He was interested in the function of "truth", that is, the possibility to judge whether beliefs and theories should be accepted or rejected. For this reason, I suggest that a specific Sellarsian pragmatist notion of truth, which understands truth as semantic assertability, serves Kuhn's interests well.

The evolutionary development of science then also highlights the relation between coherence and truth. Does truth entail coherence? Could an incoherent set be true? My conclusion is that even if science does not produce a coherent global true picture of the world, it can generate a large but unified set of "truths" entailed by scientific disciplines locally. This is not to say that the sciences as a whole form an inconsistent set in which beliefs contradict each other, although that too is possible in an extremely large set of scientific beliefs. This is to say that the set of beliefs does not cohere well; that is, their inferential connections are weak or non-existential. The assumption that must be rejected is the idea of unified science and an internally cohesive system of knowledge.

11.2 The Negative Rationale: The Failure of the Teleological View

In the postscript of SSR, Kuhn makes the first explicit appeal for the conceptualization of scientific development as an evolutionary process and not as a teleological approach towards a(ny) goal. It is notable that this appeal is made in contrast to a more customary account, according to which "science [is] the one enterprise that draws constantly nearer to some goal set by nature in advance" (SSR-2, 170). In order to understand Kuhn's case for the evolutionary process of science, it is useful to first consider what is wrong with the teleological approach.

The key idea in the teleological view is that the developmental process has a telos: a goal. Kuhn compares the required revolution regarding the nature of scientific progress to the one effected by Darwin in evolutionary biology. Kuhn's reading of Darwin's revolution in biology is that the idea of evolution was resisted, not because of the lack of evidence, but because it was so difficult to abandon the thought that the evolutionary process was goal-directed: "The 'idea' of man and of the contemporary flora and fauna was thought to have been present from the first creation of life, perhaps in the mind of God. That idea or plan had provided the direction and the guiding force to the entire process" (SSR-2, 170–171). Remarkably, it was not only theological anti-Darwinists who believed in God's plan but also pre-Darwinian evolutionary theorists, such as Lamarck, Chambers, Spencer and the German *Naturphilosophen*.

The goal-directed, teleological, approach could naturally also be something other than truth-directed as long as the goal is determined in advance, and the goal can naturally be set by something other than a god. Referring to the developmental process of science sketched out in SSR, Kuhn writes that "nothing that has been or will be said makes it a process of evolution *toward* anything" (170). Nevertheless, it is the teleological approach to the *truth* that Kuhn has in mind and that he denies. He asks specifically whether it "really help[s] to imagine that there is some one full, objective true account of nature and that the proper measure of scientific achievement is the extent to which it brings us closer to that ultimate goal" (SSR-2, 170). Further and more directly, Kuhn writes that, quite like biological evolution, "the entire [developmental] process [of scientific knowledge] may have occurred ... without the benefit of a set goal, a permanent scientific truth, of which each stage in the development of scientific knowledge is a better exemplar" (SSR-2, 171–172). This raises the question that, if it is not the preordained truth of the world that drives the scientific process, what is it? This is the question to which Kuhn's evolutionary model provides an answer.

In the language of contemporary history and philosophy of science, Kuhn thus wished to reject convergent realism, which states that science converges upon truth over time. Contemporary convergent realism is a non-theological version of teleologism. While probably no contemporary convergent realist believes that there is literally a predetermined plan for the sciences, and even less that there is a being that has set this plan for them, they still believe that nature or reality determines what the scientific theories will be in the long run. In this sense, the goal, which is truth or one, and only one, correct view of the world, is set in advance.

Let us pause here and consider in more detail what the teleological approach, in which the telos is truth, entails. For Kuhn, it presupposes the following three premises:

(a) the sciences together form, or at least strive towards, a unified view of the world;
(b) there is continuity in theory transitions in history (towards this true view of the world);
(c) truth is correspondence to the mind-independent world.

Kuhn thinks that all these premises are false, which makes the teleological view untenable and paves the way for an evolutionary account. Now, the quotations from SSR make the entailment, and the denials, of (a) and (b) clear. The premise (a) is rejected directly, that is, the rejection of the idea that "there is some one full, objective true account of nature" abandons the hope of acquiring a unified scientific world view. The "full, objective true account of nature" ought arguably to be a unified view of multiple scientific disciplines. Later Kuhn's definition of the evolutionary development of science as a process of speciation then provides a philosophical explanation for this earlier scepticism in seed in SSR regarding the possibility to achieve a unified scientific view of the world. The evolutionary process leads to more variety instead of more unity. The premise (b) is then rejected indirectly. To assume that there is a progress that brings us "closer to that ultimate goal" presupposes that there is some form of continuity in this process. For how else could there be a movement towards one goal? Of course, there could be gaps and jumps, but in the long run of this historical trajectory towards the "ultimate goal", the trajectory must show continuity.

But why deny that the sciences develop continuously in history? In "Logic of Discovery or Psychology of Research", published in 1970, the same year as the Postscript of SSR, Kuhn bluntly states that "we cannot recognise progress" towards truth (ET, 289). In "Metaphor in Science", from 1977, the answer is similarly clear: "I see no historical evidence for a process of zeroing in" on the world's real joints (RSS, 206). These excerpts make it clear that the early Kuhn argued that the history of science is not a continuous process, because empirical evidence in the historical record does not support this view. It then follows that if the history of science contains various kinds of ruptures, it cannot gradually converge on any goal. For this reason, convergent realism is rejected on empirical grounds, as being incompatible with the historical record.

However, Kuhn's view changed at some point in the time between his early philosophical texts and his characterizations of the evolutionary development of science in his late writings. Tellingly, Kuhn says that "many of the most central conclusions we drew from the historical record can be derived instead from *first principles*" (RSS, 112). He also writes that from "what I shall call the historical *perspective*, one can reach many of the central conclusions we drew with scarcely a glance at the historical record itself" (RSS, 111 my emphasis). He thought that he and his generation of historical philosophers of science overemphasized the empirical component in attempts to form an adequate image of science: "[T]he evolutionary epistemology need not be a naturalized one" (RSS, 95). These are striking statements from the philosopher who has often been perceived as a naturalist and become famous for demanding historical accountability for our thinking about science. Brad Wray correctly states that the early Kuhn's reliance on evidence and empirical results was replaced by the later Kuhn's philosophical perspective on science (2011, 84–85, 87, 95–96), which Kuhn calls, slightly confusingly, the "historical perspective". The historical perspective is in turn the same as the evolutionary or developmental view of science. It outlines a particular approach for how to study and approach science and defines what scientific knowledge is like: fundamentally, science is a process rather than a static body of beliefs, and all explanations must respect this developmental perspective.

How about the rejection of the presupposition (c) that truth must be conceived of as correspondence to the mind-independent nature? On a number of occasions, Kuhn expressed his dissatisfaction with the correspondence theory of truth. In essence, it is the evolutionary or the developmental perspective that challenges the correspondence theory of truth: "[W]hat is fundamentally at stake is ... the correspondence theory of truth, the notion that the goal, when evaluating scientific laws or theories, is to determine whether or not they correspond to an external, mind-independent world" (RSS, 95). Now, Kuhn's fundamental reason to reject the correspondence theory cannot be spelled out before the evolutionary model of scientific development is studied in more detail. It is also worth noting that Kuhn does not wish to reject the idea of truth itself: "[W]hat replaces [the correspondence theory] will still require a strong conception of truth" (RSS, 91). Similarly, in a lecture at the turn of the 1990s, Kuhn writes that the point in rejecting the correspondence theory is that "no sense can be made of the notion of reality as it has *ordinarily functioned* in philosophy of science" (RSS, 115). It is therefore necessary to return to the concept of truth after it is clearer what the evolutionary view of science entails, and ask then how the notion of truth should function in science.

11.3 The Positive Rationale: The Explanatory Goodness of the Evolutionary Account

The positive rationale for the evolutionary view is that it succeeds in providing a historically plausible account of scientific development unlike the teleological account. One way to express this is to say that Kuhn's evolutionary epistemology provides a philosophical explanation for the early Kuhn's historiographical realization that science is discontinuous and incoherent over time. Similarly, Marcum thinks that Kuhn's evolutionary epistemology emerged as a reaction to SSR and to the conclusions drawn in it about the historical-philosophical development of science (2015b, 53). I will first discuss scientists' attempts to create a coherent scientific world view, which is however frustrated by their need to specialize and develop more accurate tools and theories. Kuhn's evolutionary view emphasizes that specialization is an essential prerequisite for the progress of science, which then also explains why both (i) unity and coherence and (ii) goal-directionality and continuity are *impossible* to achieve in the long run.

The evolutionary development of science means proliferation of scientific disciplines with all accompanying conceptual, communicational and sociological features. It is worth quoting Kuhn at length:

> After a revolution there are usually (perhaps always) more cognitive specialties or fields of knowledge than there were before. Either a new branch has split off from the parent trunk, as scientific specialties have repeatedly split off from philosophy and from medicine. Or else a new specialty has been born at an area of apparent overlap between two pre-existing specialties, as occurred, for example, in the cases of physical chemistry and molecular biology ... the new shoot ... becomes one more separate specialty, gradually, acquiring its own new specialists' journals, a new professional society and often also new university chairs, laboratories, and even departments. Over time a diagram of the evolution of scientific fields, specialties, and subspecialties comes to look strikingly like a layman's diagram for a biological evolutionary tree. (RSS, 99)[4]

Beyond this, Kuhn provides only a sketchy description of speciation in the history of science: Mathematics included astronomy, optics, mechanics,

---

[4] Politi suggests that this analogy should be qualified as "in some cases, though not necessarily every time, the creation of a new discipline does not change just the number of the branches of the tree, but the very structure of the tree" (2018, 2290).

geography and music in antiquity; natural philosophy included speculative chemistry and physics prior to the seventeenth century; and medicine included the early biological sciences. He also refers to his personal experience in the proliferation of different fields of life sciences and physics and their departments, journals and societies. "What results is the elaborate and rather ramshackle structure of separate fields, specialties, and subspecialties within which the business of producing scientific knowledge is carried forward" (RSS, 117).[5]

Politi emphasizes the shared features between the later Kuhn's accounts of scientific revolution and specialization. He suggests that both are *driven* by what he calls "specialty incommensurability", which includes both semantic and methodological dimensions (2018, 2077–2080). However, it is not clear whether and when incommensurability leads to speciation, rather than triggers one. One thing is that during the process of speciation, communication is possible, and the development of more full-fledged vocabulary takes place only after a split. Furthermore, it seems that the prespeciation and prerevolution moments are appraisal disagreements regarding the nature of scientific ontologies, problems and promise of rival research trajectories. The appraisal disagreements are also bound to accelerate and deepen speciation (2018, 2282–2283).

Regarding the development of the sciences on the whole, the evolutionary model has two important consequences. One is that it destroys any hope that the sciences would generate a unified world view. The other is that it frustrates any attempts to find overall continuity in science. The direction is rather towards further fragmentation. That is, the evolutionary model provides a principled explanation for the incoherence and the fragmentation of scientific disciplines observed in the historical record. Compare how Kuhn comments on an important consequence of the evolutionary development of science:

> To anyone who values the unity of knowledge, this aspect of specialization – lexical and taxonomic divergence, with consequent limitations on communication – is a condition to be deplored. But such unity may be in principle an unattainable goal, and its energetic pursuit might well place the growth of knowledge at risk. Lexical diversity and the principled limit it imposes on communication may be the isolating mechanism required for the development of knowledge. (RSS, 99)

---

[5] See Wray (2011, 122–127) for a very useful reconceptualization of the mechanisms of Kuhn's model of scientific speciation and two illustrations of it beyond Kuhn (Wray 2011, 127–130).

The basic inclination of scientists is to move from incoherence towards coherence, as seen in "the energetic pursuit towards the unity of knowledge", but these efforts are necessarily frustrated because of the "isolating mechanisms" required for the development of knowledge. Kuhn writes that

> [w]e know little about historical changes in the unity of the sciences. Despite occasional spectacular successes, communication across the boundaries between scientific specialties becomes worse and worse. Does the number of incompatible viewpoints employed by the increasing number of communities of specialists grow with time? Unity in the sciences is clearly a value for scientists, but for what will they give it up? (ET, 289)

Kuhn notes that in a revolutionary change, one must either live with incoherence or revise several assumptions and interrelated generalizations. In other words, only the situation before and after speciation may provide a "coherent account of nature" (RSS, 29). When an incremental addition or small change is not possible, a more holistic revision is required. For example, "in the case of Aristotelian physics, one cannot simply discover that a vacuum is possible or that motion is a state, not a change of state. An integrated picture of several aspects of nature has to be changed at the same time" (RSS, 29). The result is thus that while scientists aspire to make their theories and belief systems more unified and coherent, the desire for increased specialization and accuracy results, first, in revising scientific theories and taxonomies, and in the long run, in the speciation and the fragmentation of the sciences.[6]

Kuhn's argument is that scientific fields are, in effect, practices that attempt to create more specialized and accurate knowledge of the world, which is reflected in the incoherence and defragmentation of the scientific world view and of the sciences as a whole. In brief, the evolutionary account explains these observable historical manifestations of science.

In comparison to Kuhn's early writings, his view of the two central concepts, revolution and incommensurability, changes in his later career. While the Kuhn of SSR is the well-known philosopher of sudden and

---

[6] One might well ask what "coherence" then means. And what does it mean to say that the scientific world view becomes more incoherent over time? Now, "coherence" can be understood as the requirement that a belief system is consistent; that consistency, the number of inferential relations between beliefs and other components of a belief system, must be maximised (and conversely, the number of unconnected beliefs and components minimized); and that the number of anomalies must be minimized (see Bonjour 1985, chapter 5; see also Kuukkanen 2007 and Section 11.5 below; Sankey (forthcoming)).

wholesale revolutions, the mature Kuhn's "revolution" signifies a transformation that requires a local taxonomic change and means untranslatability between two lexical taxonomies.[7] Furthermore, incommensurability plays a special and an explicitly positive role. As Kuhn put it in the paper, "The Road since *Structure*": "incommensurability is far from being the threat to rational evaluation of truth claims. . . . [It is] rather what is needed, within a developmental perspective, to restore some badly needed bite to the whole notion of cognitive evaluation" (RSS, 91). The evolutionary development of the sciences, or the speciation of separate disciplines, also means specialization with regard to language. Disciplines create their own languages and taxonomies suited to them. The limitation that specialization imposes on communication and a community is "the necessary price of increasing powerful tools" (RSS, 98). He compares "a community of intercommunicating specialists" to "a reproductively isolated population", bound together by a shared gene pool (RSS, 98). Wray has aptly compared the function of incommensurability in the sciences to how mountain ranges or wider waterways divide ecological niches and enable specialization (Wray 2011, 75). Politi (2018, 2280–2282) emphasizes that the communication barrier is partial at most. While communication across disciplines may be difficult, it is not typically manifested as a total lack of communication. Indeed, one way in which communication is possible even when a translation is not is to become bilingual (RSS, 77). In any case, incommensurability with its communication difficulties provides the needed space for scientific disciplines to develop in their own directions.

In sum, Kuhn's evolutionary account of scientific development makes it clear why

> (i) there *cannot be unity and coherence* of the sciences and their world views: speciation leads to increased specialization and resulting incommensurability between scientific disciplines;
> (ii) there *cannot be continuity and direction towards a goal* in the development of the sciences: speciation results in the increasing disintegration of disciplines and the multidirectional fragmentation of the tree of the sciences.

These are consequences from Kuhn's philosophical "historical perspective", that is, evolutionary epistemology. In addition, from the early Kuhn's argument, it is also clear why

---

[7] Kuhn defines this via the "no-overlap principle", according to which "no two kind terms, no two terms with the kind label, may overlap in their referents unless they are related as species to genus. There are no dogs that are also cats, no gold rings that are also silver rings, and so on" (RSS, 92).

(iii) there *is de facto no continuity and directionality towards a goal* in the development of science: the historical record does not support this assumption.

11.4 The Role of "Niche" in Scientific Practice

What is the fundamental reason for Kuhn's disavowal of the correspondence theory of truth? It may not be obvious at first sight. Even if science would develop discontinuously and scientific theories in total would not produce a coherent world view, the correspondence theory cannot be discounted as a viable theory. It may be useless or inapplicable under these circumstances, but not untenable as a theory of truth, as an expression of what it means to say that something "is true". It is necessary to say more about the correspondence theory of truth and its rejection in the context of Kuhn's philosophy. Understanding the reasons for Kuhn's rejection of the correspondence theory of truth also makes the most profound positive contribution of the developmental view apparent: how it explains scientific change.

Two metaphors are striking in Kuhn's depictions of the nature of scientific change. Consider the following words in "Trouble with the Historical Philosophy of Science": "Scientific development is like Darwinian evolution, a process *driven* from behind rather than *pulled* toward some fixed goal to which it grows ever closer" (RSS, 115; my emphasis). In PW, the same view is expressed:

> For the developmental movement [scientific hypotheses] ... are to be compared with the body of belief they aim to replace. One view sees science as *pulled* from ahead by the reality which a body of true belief would capture, the other as *pushed* from behind, away from what was actually believed in the immediate past.[8]

This raises the question of how exactly "pull"/"drive" and "push" should be understood. What could be pulling, and what pushing? "Pull" clearly refers to nature, or "reality",[9] which determines the theories that scientists generate. The choice to characterize the opposition with these terms is interesting, because if "nature" or "reality" is thought to *determine* what science and its theories are, we could easily understand this "force" pushing science. Indeed, much of pre-Kuhnian historiography of science assumed that nature determines the outcomes of science. Bruno Latour has

---

[8] PW, chapter 1 "Scientific Knowledge as Historical Product", 10, 12 and 13. My emphasis.
[9] PW, chapter 1 "Scientific Knowledge as Historical Product", 13.

provocatively described and called this view "natural realism" as opposed to "social realism", because the former assumes that the existence of objects is enough to explain why scientists agree about them (Callon and Latour 1992, 345). Kuhn's terminological choice is explained by his interest in science as a dynamic entity and by his interest in understanding what moves science: is there some force that pushes science somewhere almost by compulsion, or is there some other force that pulls it to undetermined directions without any compulsion? The function of scientists in the view of "natural realism" is to play a role of impersonal media, which channels the correct view of the world, determined by nature itself, in the long historical process. The "push" assigns a much greater role for the process itself, and in particular, for historical agents, "species", and their historical stages in their "drive" to develop science by adjusting their tools, theories and practices to fit their specific environments.

Kuhn's theory of evolutionary development builds extensively on the idea of adjustment. It is both scientists and the world in which they reside that are adjusted according to the needs of scientific practice. In science, "it is groups and group practices that constitute worlds (and are constituted by them). And the practice-in-the-world of some of those groups *is* science" (RSS, 103). The difference to a standard realist understanding is significant. Kuhn does not assume that there is a stable and given world that science approximates by its theories, but that the world in which science is located changes itself: creatures and niches evolve together. It is therefore simply impossible that science would converge on a specific view of the world, because there is no such permanent and fixed world. Kuhn still thinks that there is "something permanent, fixed, and stable ... like Kant's *Ding an sich*" (RSS, 104). But this world is "ineffable, indescribable, undiscussable, located outside space and time" (RSS, 104).[10]

This is how Kuhn described the process, in 1991:

> What replaces the one big one mind-independent world about which scientists were once said to discover the truth is the variety of niches within which the practitioners of these various specialties practice their trade. Those niches, which both create and are created by the conceptual and instrumental tools with which their inhabitants practice upon them, are as solid, real, resistant to arbitrary change as the external world was once said to be. But, unlike the so-called external world, they are not independent of mind and culture, and they do not sum to a single coherent whole of which

---

[10] Hoyningen-Huene distinguishes the "phenomenal world" from the "world-in-itself" (1993, 31–36).

we and the practitioners of all the individual specialties are inhabitants. (RSS, 120)

While Kuhn's evolutionary epistemology has been neglected to be analysed for a long time, the notion of "niche" has received even less attention. It is my view that the concept of "niche" is central for the later Kuhn's thinking. Through understanding its role, it is possible to shed light on other aspects of his philosophy as well.

Let us stop for the moment to consider what "niche" is. Needless to say, that in Kuhn it is a metaphor or analogy borrowed from biology.[11] However, Kuhn never said precisely what he meant by "niche". Could biology help here? There are several definitions of "niche" to be found in biology. Escobar and Craft (2016) write that the concept of niche has gone through four different stages prior to the modern understanding of "niche". Just to pick a couple of interesting transitions of this concept, Grinnell coined the term, defining "niche" as something like an environment to which species must adapt, such as the abiotic factors like temperature and precipitation. Elton also suggested that species may affect and shape their environment and introduced interaction with biotic factors, such as depredation and parasitism. More recent definitions have emphasized the totality of all, abiotic and biotic, factors, and the habitat that enables species to exist without migration. "The current ecological niche term refers to the environmental conditions in which a species can maintain populations in the long term without need of immigration" (Escobar and Craft 2016, 3).

---

[11] Renzi (2009) has criticized Kuhn's evolutionary analogy, because she thinks that there is a mismatch between the processes of biological evolution and scientific change. According to her, Kuhn confuses niche with ecological locality. Species create their own niches, and therefore the co-evolution is not between species and niche, but between species and the locality. However, Reydon and Hoyningen-Huene point out that Kuhn never intended a literal reading; that is, he never intended to compare the *mechanism* of biological evolution to that of scientific change (2010, 470). Kuhn used the evolutionary notion as an analogy between similarly progressive but not goal-directed processes in developmental biology and the history of science. Similarly, Marcum (2015b, 146) points out that evolutionary epistemology in Kuhn refers to the evolution of ideas, concepts and scientific theories, and not to the evolution of cognitive mechanism and capabilities underlying scientific practice. Evolutionary epistemology is thus a theory in (developmental) epistemology and not an application of developmental biology. Indeed, in the first chapter of PW, the term "Darwinian" evolutionary development in typed text is corrected by a hand-written note of "Darwinian-like". This further supports the idea that Kuhn meant this comparison as analogical; naturally, "analogy" is already mentioned explicitly in SSR-2 (172). However, Marcum criticizes Kuhn for relying only on a single, gradual, tempo of evolutionary scientific progress, arguing that there are in fact several tempos (2012b; 2015b, 145, 172). See also Wray (2011) for criticism of the naturalistic and literal reading of the evolutionary analogy (81–82, 103).

For the purposes of this essay, the most important element in this *analogy* is that niches both mould their practitioners and are moulded by their practitioners. It is a two-way process. This is quite as Kuhn emphasized in the quotation: "[N]iches . . . both create and are created by the conceptual and instrumental tools with which their inhabitants practice upon them" (RSS, 120). Because of this, there is no fixed world, no stable "truth-maker", that would make theories true. Both the truth-bearer, scientific theories and other propositional entities, and the truth-maker, the world of scientists, evolve together. This is the fundamental reason why the correspondence theory in the standard realist sense is inapplicable in a Kuhnian framework.[12]

Furthermore, the concept of niche makes the nature of the "push" in scientific transitions understandable. In scientific transitions, scientific communities attempt to adapt and find new solutions to the problems that they inherit. Nature does not dictate what the world of scientists' theories must be. Instead, scientists must find their own adaptations, and they are constrained by the historical developmental stage and the need to find solutions to the problems inherited. Consider how Kuhn describes the historical process from the developmental/historical perspective. The evolutionary view outlines an extended *process*, an example of which is the discovery of vacuum:

> It [the developmental view of science] recounts a series of closely linked stages, between which the cognitive distances are small: from the impossibility of a vacuum; to the impossibility of an extended vacuum; to the demonstration that man can overcome nature's abhorrence of a vacuum (the water column's breaking under its own weight); to the mercury version of the water experiment, and finally, to Torricelli's hypothesis of the sea-of-air.[13]

This is an excellent illustration of the historical chain of reasoning, as all new stages in the developmental process are preconditioned by and follow from the earlier stages. They take the earlier context as given and attempt to find better solutions and tools. Wray talks about the "tradition-bound nature of science": "changes in science are best understood as responses to

---

[12] Naturally, the correspondence theory could be used without the assumption that the world is mind-independent insofar as there is a relation of correspondence somehow between a truth-bearer and truth-maker. Nevertheless, the correspondence theory and its variants like Tarski's are arguably the favoured theories of realists, who assume that the truth-maker is something mind-independent. For truth-bearers and truth-makers, see MacBride (2019).
[13] PW, chapter 1: "Scientific Knowledge as Historical Product", 12.

existing problems, not as attempts to get at a description of the world as it *really* is" (Wray 2011, 113–114).

This analogy of "niche" can be pushed further to make sense of the idea, expressed in the words of SSR, that "though the world does not change with a change of paradigm, the scientist afterward works in a different world" (SSR-2, 121). After the emergence of a new discipline, its practices, lexicon, problems and explanations, that is, the whole "niche" inhabited, are different, although there is no change in the world in the Kantian sense of *Ding an sich*. A case in point is the emergence of the concept of virus and the field of virology. Bacteriologists, chemists and virologists inhabit phenomenally different worlds with their tools, concepts and explanations, despite the fact that, noumenally, they inhabit the "same" old world (see Wray 2011, 127–130).

11.5 Truth and Coherence

There appears to be two incompatible tenets in Kuhn's philosophy. He wishes not to abandon truth as a concept. Nevertheless, Kuhn maintains that the sciences form an incoherent world view. Can an incoherent world view be true? Is Kuhn committed to incompatible predicates of knowledge: truth and incoherence? What does "true" mean if these two are incompatible?

Nicholas Rescher writes that "the conception that all truths form one comprehensive and cohesive system in which everything has its logically appropriate place, and in which the interrelationships among truths are made duly manifest, is one of the many fundamental ideas contributed to the intellectual heritage of the West by the ancient Greeks" (Rescher 1982, 168). He calls this conception "truth as a system". In addition to Parmenides and Aristotle, Rescher identifies Kant as an advocate of this idea. This looks like an expression of the unity of science or (scientific) knowledge in another form. Rescher postulates that coherence entails three features: comprehensiveness, consistency and cohesiveness, which is also called unity. A condition of cohesiveness says that "the true propositions form one tightly knit unit, a set each element of which stands in logical interlinkage with others so that the whole forms a comprehensively connected and unified network" (Rescher 1982, 173).

This requirement of coherence is typically applied to "our knowledge" of the world and not to the world as such. The idea is thus that the concepts of coherence and incoherence are not applicable to nature but to our descriptions of it. However, supposing that Kuhn sticks to the notion of

Truth, Incoherence, and the Evolution of Science 217

truth, his evolutionary model allows that the scientific world view contains a large set of true things said about the world in different scientific specialties, which do not form an internally coherent set. This is because true propositions of the world in one scientific specialty may be so far removed from the propositions of another that their inferential connections are poor or lacking. Although they may not necessarily contradict each other, they may thus be unrelated and relatively irrelevant to each other as a consequence of being mutually incommensurable in the sense of the later Kuhn. Kuhn's evolutionary model clearly questions the idea of "truth as a system", and specifically its assumption that all truths form a cohesive and tight-knit unit.

From a global perspective, Kuhn's evolutionary model is pluralistic. Scientific knowledge takes diverse and undetermined forms. Kuhn moves from the global concerns of scientific knowledge in his early writings to the local concerns of it in his later writings. While the notion of coherence is not, or only very poorly, applicable to all sciences and to all scientific knowledge, it is applicable to local forms of knowledge. Each specialty develops theories that are both as consistent with each other and contain as high number of explanatory relations between them as possible. Each scientific speciality as a specific pocket of knowledge thus attempts to form as tight-knit explanations of the world as possible.

Furthermore, Kuhn's evolutionary model can be said to be yet more deeply pluralistic in that pluralism applies not only to scientific knowledge, to the descriptions of the world, but also to the world itself. The concept of niche entails that both scientists and the world itself are modified in the historical process. On the level of scientific research, the developmental perspective does not imply that there is one harmonious world or that all the worlds of the sciences can be reduced to one. These worlds inhabited by scientists are irreducibly different and incommensurable. This is a significant difference to an alternative model in which the incoherence of scientific knowledge is an intermediate stage on the way towards overall coherence, as in "truth as a system", gradually reflecting more accurately the world that is itself harmonious. Kuhn's world, or rather worlds, are not required or expected to be in harmony.

What is the meaning of "truth" then for Kuhn in all this? It is now clear from the above that Kuhn's first point is negative. Kuhn rejects the correspondence theory in the sense that it requires correspondence to any frame-independent, or lexicon-independent, reality. It is symptomatic of Kuhn's discussion on truth that he sees it as essential that "truth" enables assessing what to accept and reject. For example, after prompting us to reject the correspondence theory of truth in "Road since *Structure*" he moves on to pragmatic and epistemic considerations of *evaluation*. One might be tempted to call this an epistemic notion of truth in that the role of truth is inherently tied with a prospect for evaluation. However, Kuhn did not favour an epistemic theory of truth. He suggests that the correspondence notion of truth is replaced by a *redundancy* theory of truth, which relies on the laws of logic and specifically on the law of non-contradiction. But rather than understanding truth as a semantic notion, as epistemologists and metaphysicians normally do, Kuhn again highlights epistemic considerations: "[T]he essential function of the concept of truth is to require choice between acceptance and rejection of a statement or theory in the face of evidence by all" (RSS, 99).

I now wish to explore Kuhn's relation to pragmatism and the pragmatist conceptions of truth. I argue that although Kuhn rejected the pragmatist understanding of truth, the pragmatist approach to knowledge in general fits his developmental perspective and manages to explain how there can be many "truths" in an incoherent world view of the sciences. John Dewey famously proposed that "knowledge" be replaced by "warranted assertability", although he also approved of the Peircian idea of truth as the opinion that is agreed upon by all in the ideal epistemic circumstances.[14]

What is the problem with the pragmatist approach, according to Kuhn? Kuhn writes that "the basic difficulty with pragmatic conceptions of truth is their failure to explain the need to choose between incompatible theories and laws". He continues in an apparent indirect reference to Dewey that "a second formulation which equates truth with warranted assertability is equally incapable of forcing choice: there are often excellent evidential reasons for asserting each of two incompatible laws or theories. No effective scientific community can endure such inconsistency for long".[15]

Here we have another expression for the view that scientists aspire for internal coherence by attempting to get rid of incompatibility, only to be forced to increase the incoherence of the sciences through the diversification of the scientific fields in total.[16] In actuality, the Deweyan kind of

---

[14] Dewey writes that "when knowledge is taken as a general abstract term related to inquiry in the abstract, it means 'warranted assertability'" (1938, 9). He also writes that "the best definition of truth from the logical standpoint which is known to me is that of Peirce: 'The opinion which is fated to be ultimately agreed to by all who investigate is what we mean by the truth, and the object represented by this opinion is the real'" (1938, 345).

[15] PW, chapter 1, "Scientific Knowledge as Historical Product", 22.

[16] Kuhn's quote continues: "If [inconsistency] persists, the affected community regularly restores the law of non-contradiction by bifurcation, one group employing one theory, another an incompatible one. But bifurcation is a last resort. Scientific development appears to depend upon scientists'

pragmatist theory is ideally suited for Kuhn's purposes to force a choice between competing statements and theories, which he has thus set as the most important criterion for the notion of truth. Kuhn has provided a set of epistemic criteria, which can be used to assess rival claims and theories, and these criteria of evaluation are normally enough for the "cognitive bite of science".[17] To simplify somewhat, the theory or claim that scores highest in epistemic values also acquires the highest warrant and should be accepted, and its rivals rejected. If this, forcing a rational choice, is what Kuhn wished from a theory of truth, the pragmatist approach delivers. Indeed, he writes positively about "rational assertability", and specifically the values of compatibility and the law of non-contradiction, as a way to judge whether a statement is true or false. After it has been determined whether a particular "lexicon-dependent" statement is a "candidate for true/false", we must ask, "[I]s the statement rationally assertible? To that question ... the answer is properly found by something like the normal rules of evidence" (RSS, 99). It is another matter whether "warranted assertability" should replace truth or be merely understood as a criteriological and a practical approach to judge what is most likely true. The main issue is that there are standards that enable the acceptance and rejection of theories and laws.

Neopragmatism in the form of Wilfrid Sellars and Robert Brandom can be used to develop the Kuhnian idea of "truth" further. Sellars was interested in the functions of linguistic entities including philosophical terms. According to Sellars's theory, the meaning of an expression is the role it plays in a language. The strategy from a Sellarsian metalinguistic perspective is to look for suitable linguistic roles so that the "statements about linguistic roles are reducible to statements involving no abstract singular terms" (Sellars 2007, 157). From a Sellarsian perspective, it is natural to reject the idea that truth refers to an abstract entity, which accords well with the rejection of the correspondence theory of truth in Kuhn's writing. This is to say, we do not need a substantial theory of truth in which "truth" is a predicate in sentences and refers to some kind of

---

resisting the easy fragmentation of their communities" (PW, chapter 1, "Scientific Knowledge as Historical Product" 22).

[17] Interestingly, the *standards* of evaluation, and thus rationality, are not relative. In addition to mentioning the same set of epistemic values in a number of texts, Kuhn says that the criteria like "accuracy, scope, simplicity, and so on . . . are themselves *necessarily permanent*, for abandoning them would be abandoning science together with the knowledge which scientific developments bring" (RSS, 252; my emphasis).

abstract property. Now the key question is: What is the linguistic role of "truth" that could be used to substitute the abstract concept of truth?

Sellars's view is that for a proposition to be true, it means that it is assertible: "True is the same as semantically assertible" (Sellars 1967, 101). Brandom's interpretation of Sellars's position is that talk about truth is "misleading talk about what one is *doing* in saying something in the sense of making a statement: the use of 'truth' is to be understood in terms of the platitude that asserting is taking true" (Brandom 2015, 258). The Sellarsian view of truth is a form of deflationarism in something like the following way. The proper use of "true" enables one to make the following inference: from "Bob is five feet tall is true" to "Bob is five feet tall." In order for this to work, it is necessary to understand the kinds of inferences that are warranted.

Sellars's idea is that the use of language always takes place in a frame and in the space of reasons stipulated by a specific language game and its world view. The idea of frame-dependent inferences well serves Kuhn's request that "truth" must be able to discriminate between epistemic candidates. "Truth" thus plays a special social-discursive role and its *correct application* in sentences means that the inferences taken are appropriate. The focus in judging whether an inference is warranted must be determined on the basis of the inferential specifics of a framework. Nevertheless, it is difficult to outline a general theory of correct inferences because we are dealing with material inferences whose validity depends on non-logical vocabularies.

I finish this section on truth and coherence with an illustration of legitimate or warranted and illegitimate or unwarranted semantically assertible inferences in a frame of pre-Copernican and Copernican astronomy (cf. SSR-2, 115). A pre-Copernican could say that "the yellow object in the sky is a planet is true" (pointing to the Moon), dropping "true", and inferring therefore that "the yellow object in the sky is a planet". In the pre-Copernican framework this inference is legitimate and semantically assertible. In a Copernican framework, this reasoning is indicative of the set of accepted but unwarranted beliefs. A Copernican would say something like "the yellow object in the sky is a satellite is true" inferring that "the yellow object in the sky is a satellite". Although the notion of "truth" hardly adds anything substantial, this is one way in which it can be kept as a (more or less redundant) linguistic device, quite as Kuhn required: it indicates an approval of the stated, and by extension, correct inferences maintaining truth-values constant.

In conclusion, this pragmatist notion of truth allows one to discriminate between acceptable and unacceptable inferences. It also sheds light on the frameworks and on how their legitimate rules of reasoning vary in the history of science. Like Kuhn said, there is no Archimedean platform. Finally, when this idea of truth is applied synchronically to different coexisting frameworks or lexicons of science, it becomes understandable how there can simultaneously be different and distinct truths, that is, semantically assertible statements, in disciplines, given the incoherence of the overall system of the sciences.

11.6 Conclusion

This essay has examined Kuhn's reasons for proposing that science progresses *like* evolution in developmental biology. My suggestion is that Kuhn had both positive and negative rationales. The negative rationale is the difficulties that the rival account, the teleological model, is faced with in attempting to provide an account of scientific development. Kuhn argues that the historical record does not support the goal-oriented view, where the goal is truth. The positive rationale is that the evolutionary account manages to explain the dynamics of science evident in the historical record. Scientific fields specialize and develop ever better tools to cope with the world, resulting in the historically documentable fragmentation and incoherence of scientific disciplines and the scientific world view. Because scientific practitioners adapt to their environments, and modify their environments, and because the world of scientific practice changes as a result, there is no permanent fixed world to which the correspondence theory of truth can be applied in the history of science. The primary function for "truth" in Kuhn is that it enables making choices between incompatible theories and laws. My suggestion is that the pragmatist notion of warranted or semantic assertability serves these purposes excellently.

Acknowledgement

I thank James Marcum and Brad Wray for their invaluable feedback and comments on an earlier draft of this paper.

CHAPTER 12

Reassessing Kuhn's Theoretical Monism
Addressing the Pluralists' Challenge

K. Brad Wray

12.1 Introduction

Scientific specialties play a central role in Kuhn's philosophy of science. They are the key unit of analysis in his theory of scientific change. It is scientific specialties, after all, that sustain normal science research traditions. And it is scientific specialties that undergo radical changes of theory. Kuhn believed that in the normal course of development in the natural sciences, scientific specialties are characterized by a commitment on the part of the members of a specialty to a single dominant theory. That is, scientific specialties in their normal scientific phases are characterized by theoretical monism. This is what makes scientists working in the natural sciences so efficient in realizing their epistemic goals. It is also what distinguishes them from the social sciences. The sort of consensus that characterizes the community of scientists working in a specialty during periods of normal science is thus central to Kuhn's explanation for the success of science.

Recent work in the philosophy of scientific practice has raised some serious questions about the extent to which there is or needs to be consensus in science, thus challenging a key dimension of Kuhn's view. This is, to some extent, a reaction to the traditional focus on theories in philosophy of science.[1] Hasok Chang has been a leader in this project, focusing our attention on the importance of the practices constitutive of science. At its core, Chang's project is normative. He wants us to see that there are

---

[1] Ian Hacking was one of the first to raise the concern that philosophers of science are too theory-centered in their analyses. In Hacking's words, "philosophers of science constantly discuss theories and representation of reality, but say almost nothing about experiment, technology, or the use of knowledge to alter the world. This is odd, because 'experimental method' used to be just another name for scientific method" (see Hacking 1983, 149).

many benefits to pluralism, benefits that are threatened when fields are characterized by theoretical monism, the acceptance of a single dominant theory (see also Ankeny and Leonelli 2016).

On the one hand, this is a welcome development in philosophy of science. Our understanding of science is far less idealized than it was in the past. Indeed, to some extent, this is just a further, and natural, development of the project begun by the historical school in philosophy of science (on the historical school, see RSS, 91). The proponents of the historical school, after all, insisted on examining the actual history of science. On the other hand, I am inclined to think that Chang, and other pluralists, for that matter, may be overcorrecting. The call for pluralism seems to overlook the importance of a certain type of consensus in science, a type of consensus to which Kuhn drew our attention. Furthermore, I think that Kuhn's theoretical monism is compatible with certain types of pluralism, including the type of pluralism that is central to Chang's concerns. Hence, Chang and Kuhn are not as far apart in their views as Chang thinks.

In this chapter, I aim to show that a variety of types of pluralism can be reconciled with theoretical monism and that theoretical monism, properly understood, serves significant functions, aiding scientists in the effective pursuit of their epistemic goals. Thus, I aim to set limits to pluralism. In the end, I hope to show that Kuhn's theoretical monism not only has room for the sort of pluralism that Chang and others aims to defend. I also aim to show that it presupposes such forms of pluralism. Furthermore, I aim to clarify the exact respects in which Kuhn's account of science, which emphasizes the importance of theoretical monism, can accommodate the sort of pluralism that Chang defends.

12.2 Chang's Challenges

Let us begin by considering Chang's concern, for he is one of the most articulate and influential advocates of scientific pluralism. Chang argues that progress in science has been undermined and scientific advances have been delayed because scientists have frequently been too quick to abandon a theory for a new competitor theory (see Chang 2012b, chapter 5). It is as if the tolerance of competing theories is too much for scientists to bear. Chang, though, argues that there have been significant costs to this critical reaction to pluralism. Chang illustrates his point with a number of case studies, including what he regards as the too-hasty abandonment of the phlogiston theory in the late eighteenth century (Chang 2012b, chapter 1).

He insists that some of the insights contained in the phlogiston theory needed to be rediscovered years later. For example, according to Chang, Phlogistonists could explain "the common property of metals," something that was inexplicable "until the advent of the electron theory of metals" (Hoyningen-Huene 2008, 110 cited in Chang 2012b, 43). Were chemists more tolerant of pluralism, they may not have had to wait so long to develop an explanation of this phenomenon. On the basis of this and other case studies, Chang argues that we need to consciously encourage pluralism in science.

Traditionally, discussion of pluralism in the philosophy of science has focused exclusively on *theoretical* pluralism (see, e.g., Feyerabend 1975/1988; Kitcher 1993). Indeed, at times Chang frames the debate between monists and pluralism in terms of theories, but theoretical pluralism is not his only concern. The type of pluralism that Chang ultimately seeks to defend is somewhat broader than theoretical pluralism. Chang is concerned with what he calls "systems of practice" (see Chang 2012b, 253). The notion of a system of practice is a bit elusive. But Chang claims that systems of scientific practice "are made up of 'epistemic activities'" (Chang 2012b, 15). "An *epistemic activity* is a more-or-less coherent set of mental or physical operations that are intended to contribute to the production or improvement of knowledge in a particular way, in accordance with some discernible rules (though the rules may be unarticulated)" (see Chang 2012b, 15–16; he repeats this definition in Chang 2018, 176). It is Chang's view that we need to encourage pluralism with respect to systems of practice in science. Such a pluralism would aid in preventing the (temporary) loss of understanding that we see in the case of the phlogiston theory, discussed earlier.

Chang advocates for a policy of "conservationist pluralism." This policy states that we should "retain previously successful systems of practice for what they were (and are) still good at, and *add* new systems that will help us identify and create new realities, making new and fresh contacts with Reality" (see Chang 2018, 183; emphasis in original). Thus, were scientists to follow the policy of conservationist pluralism, there will be a proliferation of scientific practices over time. Chang's policy of conservationist pluralism is similar to Stéfanie Ruphy's call for ontological enrichment, encouraging "new styles of scientific reasoning" that bring "into being new entities that will add ontologically to the object under study" (see Ruphy 2013/2016, 31).

Chang presents two families of arguments in defence of his pluralism: (i) arguments that appeal to the benefits of epistemic tolerance and (ii) arguments that appeal to the benefits of *interaction* between competing theories or scientific practices. A number of the arguments emphasize the importance of not losing whatever truth there is in a weaker theory, one that is at risk of being discarded (see, e.g., Chang 2012b, § 5.2.2.1; § 5.2.2.2). Successor theories seldom retain *all* the successes of their immediate predecessors (see also Laudan 1984, 127–131). Thus, even though a theory is weaker than competitor theories, it may contain aspects of the truth or insights into nature that the superior competing theory does not contain. A commitment to monism will lead scientists to make tough choices, which may result in the loss of understanding regarding some phenomena. Forced to choose one of two competing theories, scientists will lose (at least temporarily) whatever understanding the weaker theory provides that the stronger theory cannot explain.

Some of the arguments for pluralism, though, emphasize the value of pitting competing theories against each other (see Chang 2012b, §5.2.3.3). By comparing competing theories with each other, scientists can sometimes notice things that they might not notice otherwise. They can evaluate competing theories along more dimensions and thus evaluate them more thoroughly than they could otherwise. Competing theories can even draw attention to different bodies of evidence, which can play a critical role in determining the strengths and weaknesses of the theories. Arguments of both these types have been made before, most famously by Paul Feyerabend (1975/1988), and much earlier, by John Stuart Mill (1859/1956).[2]

Importantly, Chang takes his argument for pluralism to be, if not an attack on, then at least, a modification of or supplement to Kuhn's view of science. As Chang explains, "the Kuhnian view of scientific dynamics would be improved by an injection of pluralism" (see Chang 2012b, 258). Chang purports to be providing the required injection. But the fact that Chang juxtaposes his pluralism with Kuhn's view of science suggests that he does recognize the significant insight that Kuhn provides in his account of science.

Let us consider the implications of Chang's injection of pluralism for Kuhn's view.

Kuhn argues that science is as productive as it is, in large part, because of the theoretical *monism* that characterizes scientific specialties in periods of

---

[2] Insofar as Philip Kitcher supports theoretical pluralism it is as a means to ensuring that a research community or specialty does not lose the superior theory, the theory that is true or closer to the truth (see Kitcher 1993).

normal science (see Kuhn SSR-4, chapters III and IV). The consensus on a single theory, Kuhn argues, ensures that those working in a specialty can take the fundamentals of a field for granted. They can assume, for example, that others are committed to the same basic ontology with respect to the entities assumed in their domain of study. Researchers can then focus on specific, concrete, often esoteric, problems that the theory seems especially fit to solve. Significantly, Kuhn not only notes that this is a pattern we see in the history of science, he positively endorses it. That is, Kuhn not only seeks to explain why science is successful, he seeks to tell us how scientific inquiry should be conducted if it is to continue to be successful. Kuhn is thus prescribing theoretical *monism*.[3] Chang, it seems, is taking issue with precisely this aspect of Kuhn's view. Chang believes that theoretical monism is incompatible or, at least, in tension, with a pluralism of scientific practices. And Chang insists that a pluralism with respect to scientific practices will benefit science. If something ought to be abandoned it would seem to be our commitment to theoretical monism.

Indeed, Chang is not the only contemporary pluralist to treat Kuhn as a foil in presenting the case for pluralism. Ruphy also explicitly positions her pluralism against Kuhn's view. Though she refers to paradigms, given the nature of her comments, she has in mind theories, that is, conceptual frameworks, the sorts of things captured in what Kuhn came to call "lexicons." Specifically, Ruphy suggests that Kuhn is mistaken when he claims "a scientist cannot work at the same time in several paradigms" (see Ruphy 2011, 1221). Furthermore, whereas Kuhn claims that "paradigms supersede one another within a discipline," Ruphy insists on the "cumulative nature of the processes of ontological enrichment and the resulting cumulativeness of foliated pluralism" (Ruphy 2011, 1221).[4] These pluralists see something pernicious in the intolerance toward pluralism that theoretical monism encourages.

---

[3] In the preface to SSR, Kuhn suggests that what distinguishes the social sciences from the natural sciences is that in the latter, but not the former, there are paradigms. The notion of paradigm that is relevant here is one that considers paradigms to be the "universally recognized scientific achievements," what he would later call "exemplars" (see SSR-4, xlii; see Wray 2011, chapter 3). After Margaret Masterman's critical analysis of his use of the term "paradigm" in SSR, Kuhn realized that "a paradigm [that is, an exemplar] is what you use when the theory isn't there" (see Kuhn 2000b 300). Even the social sciences could have paradigms in this sense. Thus, Kuhn later came to believe that what distinguishes the natural sciences from the social sciences is that in the former there is a commitment to a *single theory* at any one time. And the sort of thing he had in mind by "theory" is a lexicon (see, e.g., Kuhn 1991/2000).

[4] Ruphy does not provide clear examples of cases from the history of science to illustrate either point.

The pluralists' concern relates to a longstanding criticism of Kuhn's view. In SSR, Kuhn claimed that scientific research communities are characterized by a high degree of homogeneity and a high degree of intolerance for departures from the accepted theory. This is the dogmatic dimension of normal science that Karl Popper found so objectionable (see Popper 1970). This is what makes Popper pity the normal scientist. According to Kuhn, it is only in periods of crisis when the confidence of those working in a field is severely eroded that alternative theories are entertained.

Kuhn's insistence on theory monism struck some early critics as implausible and problematic. Some even argued that, given the central role of theoretical monism in Kuhn's account, he seemed to present us with a situation where radical theory change would be inexplicable. Larry Laudan raised this concern and presented his own reticulated model of scientific change as an alternative to Kuhn's view (see Laudan 1984, 16–18; see also Chang 2012b, 258). Laudan's reticulated model of scientific change explains how theory change, even radical theory change, can happen in a piecemeal fashion. Laudan, unlike Chang, seems to accept theoretical monism.[5] What Laudan objects to is the suggestion that theories are as inflexible as Kuhn characterizes them to be. The sort of monism that Kuhn defends seems intimately connected to a rigid form of holism that seems to make theory change impossible.

12.3 Where Is the Consensus in Science?

In the remainder of this chapter, I want to assess the extent to which these criticisms undermine Kuhn's theory of science, and especially his claims about theoretical monism, given that Chang presents his pluralism of systems of practices as a modification of Kuhn's view. I aim to show that the type of monism that Kuhn thought was necessary for normal science is compatible with a pluralism with respect to systems of practices, at least to some extent. Thus, I aim to show that Chang's pluralism of systems of practices is not such a radical departure from Kuhn's own view. In fact, reflecting on the relationship between Kuhn's theoretical monism and the

---

[5] Unlike Chang, Laudan does not provide a defense of theoretical pluralism. Rather, according to his reticulated model of theory change, the goals, methods, and theories that scientists work with can change in a piecemeal fashion, and a change in any one part is often followed by changes in the other parts (see Laudan 1984, 62–66 and 68–87; see especially Figure 4, on page 76). Laudan's point is that piecemeal changes, unlike a holistic change, can be rationally defended.

sort of pluralism that concerns Chang will aid us in better understanding Kuhn's view, especially as it developed later in his career.

I want to begin with a clarification of Kuhn's view, one that he only really made clear in his later work, as he developed his view in response to earlier criticism. In SSR, Kuhn was quite careless in articulating what exactly the relationship was between a paradigm and a theory. This proved to be quite consequential, as he ended up spending a lot of time and effort clarifying what exactly he meant by the term "paradigm" (see Wray 2011, chapter 3). Ultimately, though, he came to distinguish paradigms from theories and reserved the term "paradigm" for the exemplars that natural scientists learn and apply in the course of their research. Importantly, in the course of doing this, he was able to get a better sense of the role and nature of theories in science. As a result, Kuhn was a lot more precise later in his life about what theoretical monism entailed.

According to Kuhn, the locus of consensus in a specialty community is the scientific *lexicon* or taxonomy that scientists employ in their research (see Kuhn RSS, 90–104). That is, a scientific specialty community is characterized by its use of a specific lexicon, a scientific vocabulary that all those working in the specialty in a given time period employ in their research. This is a large part of what young scientists internalize in their training. They learn to see the world through the lens of the accepted theory, categorizing experiences in the terms of the accepted lexicon. As a consequence of their training, generally, scientists working in a specialty unreflectively and uncritically project the categories supplied by the theory onto the world. Indeed, this is what makes them effective researchers. This is why they are able to tackle research problems so efficiently. They are able to take for granted a particular worldview, the worldview that is embedded in the accepted lexicon.

This is what the various psychological experiments that Kuhn draws on are meant to illustrate. Bruner's and Postman's experimental subjects could not resist classifying the anomalous playing cards as cards from a regular deck of playing cards (see SSR-4, 62–64). Primed by their familiarity with the cards in a regular deck, the subjects fit what they saw into the categories supplied by their background beliefs.[6] Scientists are similar to these experimental subjects, except their deck of cards is the theory they learned. Unlike the scientists in training, who might still have to deliberate about

---

[6] Recently, David Kaiser has argued that these psychological experiments played a greater role in leading Kuhn to the view he developed in SSR, perhaps greater than the historical cases that he draws on to illustrate various claims in the book (see Kaiser 2016).

the phenomena before determining how to classify what they see, the trained scientist often just sees the world as the theory dictates.

Unless the scientists working in a specialty accept the same lexicon, they will not describe the same experiences in the same terms. Without a commitment to the same lexicon, communication between scientists would be challenging. Indeed, were there no commitment to the same lexicon in a scientific specialty, it is doubtful that the sciences would progress in the impressive manner that they in fact do.

Furthermore, according to Kuhn's developed view, changes of theory are just lexical changes of a very specific sort. Specifically, revolutionary changes of theory involve lexical changes that violate the no-overlap principle. They are lexical changes that reorder the relationships between kind-terms, that is, terms designating kinds of things, terms like "chemical element" and "compound," or "planet" and "star" (see RSS, 13–20). Kuhn was not especially clear in his discussions of the no-overlap principle, but what he has in mind is roughly this. In a lexicon, every kind is related to every other kind either by a relationship of exclusion or by a relationship of inclusion. Cats and dogs are an example of the first sort of relationship, and cats and mammals are an example of the second sort. No dog is a cat, and all cats are mammals. With a scientific revolution, there is a violation of these rules. The relations between classes in the new theory or lexicon do not preserve the relations between classes or kinds in the old theory.

Kuhn discussed the sorts of lexical changes introduced during the Copernican Revolution in astronomy on numerous occasions, and this example provides a vivid illustration of his point (see, e.g., RSS, 15; RSS, 94; also Wray 2011). With the change of theory, things that were previously regarded as planets, specifically, the Sun and the Moon, were no longer regarded as planets. And the Earth, something that was not regarded as a planet, came to be classified as a planet. This sort of conceptual rearrangement is characteristic of changes of theory or scientific revolutions. Not only the extensions of the key theoretical terms change, the intensions do as well. For Ptolemy a planet was a wandering star, contrasted with the fixed stars. For Copernicus, a planet is a satellite of the Sun. Clearly, with such a change in the scientific lexicon, the sorts of general claims astronomers made about planets and stars changed as well. Certain claims that the Ptolemaic lexicon implied, no longer made sense once astronomers adopted the Copernican lexicon.

Kuhn was quite insistent that the scientists in a specialty must share a lexicon. The lexicon, after all, is what makes effective communication possible. In fact, Kuhn suggests that one of the principal sources of incommensurability in science is a consequence of the fact that scientists in different fields employ terms in different ways. The same term can often have different *connotations* in different fields. As we saw earlier, the same term can even have different *denotations* in different disciplines or subfields. In SSR, for example, Kuhn discusses how physicists and chemists differ in their understanding of the term "molecule," specifically with respect to "whether a single atom of helium was or was not a molecule" (see Kuhn 1962/2012, 50). According to Kuhn, "for the chemist the atom of helium was a molecule because it behaved like one with respect to the kinetic theory of gases. For the physicist ... the helium atom was not a molecule because it displayed no molecular spectrum" (see SSR-4, 51). Importantly, the difference is with respect to how the scientists in the two fields operationalize the concept of "molecule." It is the practice of science that determines what scientists mean by a theoretical term. Indeed, as Kuhn emphasized, young scientists in training often learn the meaning of key theoretical terms in their laboratory exercises, as they apply the terms to concrete problems.

Indeed, Kuhn notes that even when scientists in different specialties or fields learn the same theory, they learn something quite different. For example, Kuhn notes that "all physical scientists ... [are] taught the laws of ... quantum mechanics, and most of them employ these laws at some point in their research or teaching" (see SSR-4, 50). But, according to Kuhn, "they do not all learn the same applications of these laws, and they are not therefore all affected in the same ways by changes in quantum-mechanical practice" (see SSR-4, 50). Hence, we should not be misled by the apparent similarities between neighbouring subfields. When scientists are trained, they still are being trained for research in a narrow specialty.

Within a field or specialty, it is imperative for the smooth operation of research that everyone uses the terms in the same way to pick out the same things. Only then can scientists communicate effectively with each other. Indeed, a revolutionary change of theory just is a change that violates this presumed consensus. In this sense, Kuhn's early remarks about gestalt shifts, religious conversions, and world changes are apt. The scientists working with a new lexicon must learn to see, and describe, the world in a new way. And exchanges with those who accept a competing theory are bound to be frustrating, as they are apt to use key theoretical terms in different ways.

12.4 A Space for Pluralism

Importantly, the sort of a consensus that Kuhn thinks is necessary for normal science is compatible with ***significant*** differences or disagreements between the scientists working in a field or specialty. In fact, such differences play an integral part in Kuhn's account of scientific change. As I will explain in detail, the differences between the ways in which individual scientists (i) interpret and apply paradigms and (ii) weigh the theoretical values in their evaluations of competing theories are crucial to the success of science. In this respect, Kuhn's account is compatible with certain types of pluralism.

First, as was stressed even in SSR, Kuhn was insistent that paradigms, that is, exemplars, are open to multiple interpretations. Indeed, it is the flexibility of paradigms, the fact that they are capable of being applied in a variety of different ways to different problems, which makes them so useful to scientists in their normal research activities (Wray 2011, 61). The creative work of the normal scientist involves applying the accepted paradigms to hitherto unsolved research problems. Kuhn was quite insistent that scientists do not work with an inflexible algorithm – a rule – that dictates unequivocally how concepts and exemplars are to be applied. Scientific research is not just the mechanical or blind application of concepts to phenomena, not even for the normal scientist. That is why Kuhn argued for the priority of paradigms (see SSR-4, chapter V). The puzzles that normal scientists confront require ingenuity and creativity.

Popper's characterization of the normal scientist as unreflective and dogmatic is thus a gross misrepresentation of what Kuhn had in mind (see Popper 1970). Popper compared the normal scientist to an "applied scientist"; he claimed that "the 'normal' scientist, as described by Kuhn, has been badly taught"; "he is the victim of indoctrination" (see Popper 1970, 53). That is not how Kuhn conceived of normal science. Though Kuhn did not agree with the Strong Programme on many issues, the proponents of the Strong Programme had a better sense of what Kuhn meant by "normal science" than Popper and many other philosophers of science (see, e.g., Barnes 1982, especially chapter 2).[7] The Strong Programmers rightly saw the flexibility inherent in normal science, the fact that different scientists can and will approach the same research problem in different ways, even though they all acknowledge the same constraints.

Second, Kuhn argued that the values that scientists appeal to when they are choosing between competing theories are also open to multiple

---

[7] But it seems that the proponents of the Strong Programme may have read Kuhn as implying a far greater degree of underdetermination in normal science than he meant to suggest (see Wray 2011, 154–160). They seem to admit fewer constraints on scientists than Kuhn does. Kuhn expressed a variety of concerns about the Strong Programme (see Wray 2015, § 12.4).

interpretations. This is a key theme in "Objectivity, Value Judgment, and Theory Choice" (see ET, 320–339). Specifically, Kuhn claims that there is no firm consensus on how the values, like simplicity and breadth of scope, that inform theory choice are to be (i) interpreted individually or (ii) weighed against each other when scientists are evaluating theories (see ET, 321–325). Because there is no firm consensus on these matters, the norms guiding scientists permit different responses, even from scientists faced with the same body of data. There is thus room for some degree of pluralism in responding to a choice between competing theories. Again, this flexibility is crucial, according to Kuhn. It is not a weakness that will be overcome with future developments in science. Rather, it is a key to understanding the success of science.

Thus, contrary to what Chang implies, Kuhn's theoretical monism is compatible, to some extent, with a pluralism of systems of practices. Kuhn not only notes that such differences exist, as a matter of fact, in every field of science, he also attributes an important constructive role to these differences in his philosophy of science. These differences are what make possible changes of theory. These differences afford scientific research communities the flexibility they need when they are confronting a crisis-threatening anomaly or set of anomalies. The flexibility afforded by paradigms and the values that inform theory choice are, according to Kuhn, sources of novelty in science. In applying the same paradigm in alternative ways, different scientists will explore different possible solutions to an outstanding research problem. They may thus be able to avert a (as-of-yet unnecessary) radical change of theory.

The fact that the values that inform theory choice do not ***determine*** scientists' evaluations of competing theories also plays a crucial role in ensuring that the various competing theories are ***developed*** when a research community is confronted with a choice between competing theories. Two scientists faced with the choice between the same two theories and, considering the same body of data, can be led to make different decisions about which theory is superior. This is, in Kuhn's words, "the community's way of distributing risk and ensuring the long-term success of its enterprise" (see SSR-4, 185–186; ET, 332; see also D'Agostino 2005; Rueger 1996). This ensures that when a new competing theory is initially developed, one that challenges the long-accepted theory, there are likely to be both: (i) some early converts to the new theory and (ii) some holdouts who continue to pursue their research in the framework supplied by the long-accepted theory. Indeed, this is what we saw during the Copernican Revolution in astronomy. Kepler and Galileo were early converts, whereas Clavius and others were holdouts (see Wray 2011, chapter 2). When research efforts in a research community are distributed, the community as a whole is less likely to (i) prematurely abandon a long-accepted, but still viable, theory, or (ii) reject a promising new theory before it has had a chance to show its superiority.

Here, we need to take Kuhn's biological metaphor quite seriously. A viable biological species needs genetic variability to be able to meet new challenges.[8] If all the members of a species were genetically identical, then a change to their environment would affect every member of the species in more or less the same way. If the change were threatening, then the species might even face extinction. Similarly, Kuhn believes that when a theory encounters anomalies, it is the differences among the scientists in the specialty, the pluralism within the research community, that provides a pool of resources for addressing the challenge. Different scientists will apply the accepted paradigms in different ways. And different scientists will evaluate competing hypotheses and theories in different ways in virtue of the different ways they interpret the values of theory choice and weigh them against each other. In this way, collectively, the scientists working in a specialty are generally able to meet the challenges that they encounter. And in this way, theories are often able to be developed so that they can successfully address anomalies.

This gets at the radical innovative dimension of Kuhn's view. Kuhn did not think that the success of science is due to many individual scientists carefully considering the data according to rigorous canons of rationality that, when applied with sufficient care, guarantees success. According to Kuhn, this traditional individualist picture of science was irreconcilable with what we learn from the history of science. Instead, he believes that the success of science is a function of the social nature of scientific inquiry and the structure of scientific specialty communities. By agreeing on the fundamentals, and working with the same scientific lexicon, the scientists in a specialty are able to concentrate their research efforts on very specific problems that the resources supplied by the accepted lexicon are especially suited to solve. Kuhn recognizes that these highly uniform research communities, unified by their shared lexicons, require flexibility if they are to function effectively. Kuhn, though, provides an account of where the flexibility comes from. In this respect, he has made room for a plurality

---

[8] As Marc Ereshefsky notes, "philosophers generally believe that species are natural kinds with essences" (see Ereshefsky 2014, 524). This species essentialism, Ereshefsky notes, is widely rejected by biologists and philosophers of biology. Kuhn's insistence on variability suggests that he is not a species essentialist.

of systems of practice within specialty communities characterized by theoretical monism.

So the differences discussed earlier provide Kuhn with the resources to address Popper's and Laudan's concern, explaining how radical theory change is possible when the accepted theory has such a comprehensive grip on those working with it. Theory change, that is, radical theory change, is possible because the scientists working in a specialty are not so uniform in their vision that change is not possible. Kuhn did not believe that scientists are forever stuck in a theory-determined vision of the world, because there is a built-in flexibility in science that manifests itself as needed, when the field is in crisis and persistent anomalies continue to resist solutions. The key to understanding Kuhn's view is to understand the delicate balance he ascribes to (i) consensus and tradition, on the one hand, and (ii) differences and creativity, on the other hand. There is a consensus with respect to the lexicon. But the lexicon can be creatively applied and developed given that scientists differ with respect to how they apply (i) the shared exemplars and (ii) the criteria of theory evaluation. This is the essential tension that characterizes science that, even before the publication of SSR, Kuhn drew attention to (see ET, 225–239).

Chang's concern can also be addressed. Contrary to what Chang implies, a certain type of pluralism is not inimical to Kuhn's view. Rather, as I have shown earlier, some forms of pluralism are an integral part of Kuhn's view. Specifically, for a research community to be able to effectively address anomalies, there must be, within the community, (i) a plurality of interpretations of the accepted exemplars and (ii) a plurality of interpretations of the values of theory choice. But there is a limit to how much pluralism can be tolerated in a research community. To be a community at all, that is, to have the sort of identity that would constitute a scientific specialty, and to function as a productive research community, scientists working in a specialty must employ the same lexicon. Otherwise, they are not even describing the world in the same way, which means they are not even working in the same world.

12.5 Conservationist Pluralism

There is another aspect to Kuhn's view of science that may address some of Chang's concerns about the need to accommodate a plurality of systems of practice. Recall Chang's policy of conservationist pluralism, which encourages us to add new systems of practice, rather than just develop new systems and discard the old. Given the policy of conservationist pluralism, scientists should tolerate different practices when researchers working in a field disagree. Furthermore, the policy encourages an ever-increasing number of practices employed in the sciences. Any research practice that continues to be effective should be retained, even though new practices have been developed and adopted.

Kuhn's view does not preclude such an attitude toward old systems of practice, despite his insistence on the importance of theoretical monism. Indeed, especially in his later writings, Kuhn provides a concrete understanding of how scientists can and do retain an ever-greater number of systems of practice, practices that may seem in tension with each other. In Kuhn's framework, often, the new systems of practice that scientists develop will find their home in new disciplines or scientific specialties. And the older systems of practice may persist in the parent discipline, where they still have a useful epistemic function.

We see something of this sort happening with respect to the various different versions of the Periodic Table of Elements that have been developed (see Scerri 2019, chapter 10; also Scerri 2007, 277–281). Some of the versions differ from each other in nontrivial ways. Scientists, though, have not felt compelled to work with only one version. Rather, as Eric Scerri notes, "many scientists tend to favour one or other particular form of the periodic table because it may be of greater use to them in their scientific work as an astronomer, a geologist, a physicist, and so on" (Scerri 2019, 129). Whether a system of practice is old or new, if it proves useful to scientists in their research it is quite likely to be retained and employed in the domain in which it works effectively. Thus, when confronted with alternative systems of practices, scientists are not forced to choose between them, retaining one and discarding the other. Instead, what initially appears to be competing systems of practice can both be retained, but employed in different domains, perhaps in different specialties or subspecialties.

This dimension of Kuhn's view, his emphasis on the proliferation of scientific specialties, became more pronounced in his later work. Scientific specialties are characterized by particular sets of practices that serve the narrow research interests of the specialty. Indeed, Kuhn suggested that the increasing success of science over time, the fact that predictions are made with greater and greater precision, for example, is a function of the increasing specialization that characterizes the growth of modern science.

Kuhn was not only committed to the view that the development of science involves a proliferation of scientific specialties over time (see RSS, 116–118), he also explicitly connects the proliferation of specialties with pluralism. He states "the sciences ... must ... be viewed as plural" (see RSS, 119). Specifically, Kuhn claims that science "should be seen as a complex but unsystematic structure of distinct specialties or species, each responsible for a different domain of phenomena, and each dedicated to changing current beliefs about its domain in ways that increase its accuracy and the other standard criteria" (RSS, 119).

Significantly, the two specialties that emerge after a field divides, the parent field and the new field, can, and most certainly will, employ different research practices. After all, the reason why the split occurs is due to the fact the two sets of scientists, working alongside each other in the parent field before the split, developed different research interests. The concepts, instruments and practices that had long served the parent field proved inadequate to the research goals of some subset of the specialty.

It is hard to see how this view of science could not accommodate the proliferation and plurality of systems of practice to which Chang draws attention. Thus, it seems that there is no need for an injection of pluralism into the Kuhnian framework. The pluralism is already there.

12.6 Concluding Remarks

In summary, I have argued that Chang's pluralism of systems of practices seems compatible with the sort of "lexical monism" which Kuhn insists is necessary for the effective operation of research communities in their normal science research phase. Chang's arguments highlight the need for the sort of flexibility that creativity feeds on. But Kuhn's view of science has room for that. In fact, Kuhn's theory of science assumes that successful research communities employ a plurality of systems of practice, including flexibility with respect to (i) the application of paradigms to scientific problems and (ii) the weighing of the theoretical values employed in theory evaluation. Furthermore, Kuhn's account of scientific specialization provides a place for Chang's policy of conservationist pluralism. As scientific specialties proliferate, systems of practices will proliferate.

Scientific specialties remain a central part of the scientific landscape. They are the locus of consensus that make effective normal scientific research practices possible, largely due to the role they play in sustaining scientific lexicons, scientific vocabularies that structure how scientists describe and conceive of reality. This sort of uniformity in the research community seems to be a prerequisite for effective research. Kuhn's theoretical monism, with its focus on the scientific specialty as the key unit of analysis, thus remains an insightful framework for understanding science, even as we aim to account for the pluralistic nature of scientific practice.⁹

Acknowledgements

I presented earlier versions of this chapter at "The Logic, Methodology, and Philosophy of Science and Technology" conference in Prague, in August 2019, and at the European Philosophy of Science Association biennial conference in Geneva in September 2019. I benefited from feedback from the audiences at both conferences. I also thank Lori Nash for feedback on an earlier draft. Finally, I thank the Centre for Science Studies, at Aarhus University, for their support. My research is supported by a Starting Grant from the Aarhus University Research Foundation – (AUFF-E-2017-FLS-7–3).

---

⁹ Despite the increase in interdisciplinary research, specialties persist in defining how individual scientists approach research problems. Indeed, Hanne Andersen argues that "interdisciplinary research can only receive epistemic justification from the established disciplines" (Andersen 2013, 7).

Bibliography

Achenbach, J. (2019). "Scientists Are Baffled: What's Up with the Universe?" *The Washington Post,* (accessed November 1, 2019). www.washingtonpost.com/gdpr-consent/?next_url=https%3a%2f%2fwww.washingtonpost.com%2fscience%2fscientists-are-baffled-whats-up-with-the-universe%2f2019%2f10%2f31%2f31fc42e4-f353-11e9-8cf0-4cc99f74d127_story.html

Agassi, J. and Jarvie, I. (2008). *A Critical Rationalist Aesthetics.* Amsterdam and New York: Rodopi.

Alexander, P. (1974). "Boyle and Locke on Primary and Secondary Qualities." *Ratio* 16: 51–67.

Allais, L. (2007). "Kant's Idealism and the Secondary Quality Analogy." *Journal of the History of Philosophy* 45, 3: 459–484.

(2015). *Manifest Reality: Kant's Idealism and His Realism.* New York: Oxford University Press.

Allison, H. (1983/2003). *Kant's Transcendental Idealism.* New Haven: Yale University Press.

Andersen, A. (2013). "A Second Essential Tension: On Tradition and Innovation in Interdisciplinary Research." *Topoi* 32: 3–8.

Andersen, H. (2000). "Learning by Ostension." *Science and Education* 9, 1: 91–106.

(2001). *On Kuhn.* Belmont: Wadsworth.

Andersen, H., Barker, P., and Chen X. (2006). *The Cognitive Structure of Scientific Revolutions.* Cambridge: Cambridge University Press.

Andresen, J. (1999). "Crisis and Kuhn." *Isis* 90, S2 (Supplement): S43–S67.

Ankeny, R. A. and Leonelli, S. (2016). "Repertoires: A Post-Kuhnian Perspective on Scientific Change and Collaborative Research." *Studies in History and Philosophy of Science* 60: 18–28.

Arabatzis, Theodore. (2005). *Representing Electrons.* Chicago: University of Chicago Press.

Ariew, R. (2006). "The Sphere of Jacques Du Chevreul: Astronomy at the University of Paris in the 1620s." In M. Feingold and V. Navarro-Brotóns (eds.), *Universities and Science in the Early Modern Period,* pp. 99–109. New York: Springer.

Averill, E. W. (1982). "The Primary-Secondary Quality Distinction." *Philosophical Review* 91: 343–362.

Ayers, M. (2011). "Primary and Secondary Qualities in Locke's Essay." In L. Nola (ed.), *Primary and Secondary Qualities: The Historical and Ongoing Debate*, pp. 136–157. Oxford: Oxford University Press.

Barker, P. (2000). "The Role of Religion in the Lutheran Response to Copernicus." In M. J. Osler (ed.), *Rethinking the Scientific Revolution*, pp. 59–88. Cambridge: Cambridge University Press.

(2001). "Incommensurability and Conceptual Change during the Copernican Revolution." In P. Hoyningen-Huene and H. Sankey (eds.), *Incommensurability and Related Matters*, pp. 241–273. Boston Studies in the Philosophy of Science, Boston: Kluwer.

(2002). "Constructing Copernicus." In P. Barker (ed.), *New Foundations in the History of Astronomy. Perspectives on Science*, 10: pp. 208–227 (published 2003). www.mitpressjournals.org/doi/abs/10.1162/106361402321147496?mobileUi=0

(2004a). "Astronomy, Providence and the Lutheran Contribution to Science." In A. J. L. Menuge (ed.), *Reading God's World: The Scientific Vocation*, pp. 157–187. St. Louis: Concordia Press.

(2004b). "How Rothmann Changed His Mind." In A. Bowen et al. (eds.), *Astronomy and Astrology from the Babylonians to Kepler. Centaurus*, 46: pp. 41–57.

(2005). "The Lutheran Contribution to the Astronomical Revolution." In J. Brooke and E. Ihsanoglu (eds.), *Religious Values and the Rise of Science in Europe*, pp. 31–62. Istanbul: Research Centre for Islamic Art History and Culture (IRCICA).

(2009). "The *Hypotyposes Orbium Coelestium* (Strasbourg, 1568)." In M. A. Granada and E. Mehl (eds.), *Nouveau Ciel Nouvelle Terre – La Révolution Copernicienne dans l'Allemagne de la Réforme (1530–1630)*, pp. 85–108. Paris: Les Belles Lettres.

(2011a). "The Cognitive Structure of Scientific Revolutions." *Erkenntnis* 75: 445–465. DOI: 10.1007/s10670-011-9333-8

(2011b). "The Reality of Peurbach's Orbs." In P. J. Boner (ed.), *Change and Continuity in Early Modern Cosmology*, pp. 7–32. New York: Springer.

(2013a). "Why Was Copernicus a Copernican?" *Metascience* 22: 1–6.

(2013b). "Albert of Brudzewo's *Little Commentary* on Peurbach's *New Theorica*." *Journal for the History of Astronomy* 44: 1–24.

Barker, P. and Goldstein, B. R. (2001). "Theological Foundations of Kepler's Astronomy." In J. H. Brooke, M. J. Osler, and J. van der Meer (eds.), *Science in Theistic Contexts: Cognitive Dimensions. Osiris*, 16: pp. 88–113.

(2003). "Patronage and the Production of *De Revolutionibus*." *Journal for the History of Astronomy* 34: 345–368.

Barker, P. and Heidarzadeh, T. (2016). "Copernicus, the Ṭūsī Couple and East-West Exchange in the Fifteenth Century." In M. A. Grenada and P. Boner (eds.), *Man and Cosmos*, pp. 19–57. Barcelona: University of Barcelona Press.

Barker, P. and Vesel, M. (2012). "Goddu's Copernicus." *Aestimatio* 9: 304–336.

Barnes, B. (1982). *T. S. Kuhn and Social Science*. New York: Columbia University Press.
  (2004). "Transcending the Discourse of Social Influences." In P. Machamer and G. Wolters (eds.), *Science, Values, and Objectivity*, chapter 5. Pittsburgh: University of Pittsburgh Press.
Batterman, R. (2002). *The Devil in the Details: Asymptotic Reasoning in Explanation, Reduction and Emergence*. Oxford: Oxford University Press.
Bernal, J. D. (1939). *The Social Function of Science*. New York: Macmillan.
Bernard of Verdun. (1961). *Tractatus super totam astrologiam*, ed. P. P. Hartmann. Werl, Wesphalia: Dietrich-Coelde-Verlag.
Bird, A. (2000). *Thomas Kuhn*. Princeton: Princeton University Press.
  (2003). "Kuhn, Nominalism, and Empiricism." *Philosophy of Science* 70, 4: 690–719.
  (2012). "The Structure of Scientific Revolutions and Its Significance: An Essay Review of the Fiftieth Anniversary Edition." *British Journal for the Philosophy of Science* 63, 4: 859–883.
Blackburn, S. (2014). "Creativity and Not-So-Dumb Luck." In E. S. Paul and S. B. Kaufman (eds.), *The Philosophy of Creativity: New Essays*, pp. 147–156. Oxford: Oxford University Press.
Bloor, D. (2016). "The Pendulum as a Social Institution: T. S. Kuhn and the Sociology of Science." In A. Blum, et al. (eds.), *Shifting Paradigms: Thomas S. Kuhn and the History of Science*, pp. 235–252. Berlin: Edition Open Access.
Blum, A., K. Gavroglu, C. Joas, and J. Renn, (eds.). (2016). *Shifting Paradigms: Thomas S. Kuhn and the History of Science*. Berlin: Edition Open Access.
Bohr, N. (1913). "On the Constitution of Atoms and Molecules." *Philosophical Magazine* 26: 1–25.
Bokulich, A. (2008). *Reexamining the Quantum-Classical Relation: Beyond Reductionism and Pluralism*. New York: Cambridge University Press.
Boner, P. (2013). *Kepler's Cosmological Synthesis: Astrology, Mechanism and the Soul*. Leiden: Brill.
Bonjour, L. (1985). *The Structure of Empirical Knowledge*. Cambridge, MA: Harvard University Press.
Boston Globe (1953). "Furry, Named in Probe, Still Teaching at Harvard." *Boston Globe*, February 26.
Brandom, R. (2015). *From Empiricism to Expressivism: Brandom Reads Sellars*. Cambridge: Harvard University Press.
Brentjes, S. (2007). "Ḥajjāj ibn Yūsuf ibn Maṭar." In T. Hockey et al. (eds.), *The Biographical Encyclopedia of Astronomers, Springer Reference*, pp. 460–461. New York: Springer, 2007, https://islamsci.mcgill.ca/RASI/BEA/Hajjaj_ibn_Yusuf_ibn_Matar_BEA.htm, (accessed 18 July 2019).
  (2009). "Cartography in Islamic Societies." In R. Kitchin and N. Thrift (eds.), *International Encyclopedia of Human Geography*, vol. 1, pp. 414–427. Oxford: Elsevier.

(2010). "The Mathematical Sciences in Safavid Iran: Questions and Perspectives." In D. Herrmann and F. Speziale (eds.), *Muslim Cultures in the Indo-Iranian World During the Early-Modern and Modern Periods*, pp. 325–402. Berlin: Klaus Schwartz.

Brentjes, S., Fidora, A. and Tischler, M. M. (2014). "Towards a New Approach to Medieval Cross-Cultural Exchanges." *Journal of Territorial and Maritime Studies* 1, 1: 9–50. DOI 10.1515/jtms-2014–0002

Brorson, S. and Andersen, H. (2001). "Stabilizing and Changing Phenomenal Worlds." *Journal for General Philosophy of Science* 32 (1): 109–129.

Burtt, E. A. (1932/1980). *The Metaphysical Foundations of Modern Physical Science*. 2nd ed. Atlantic Highlands: Humanities Press.

Callon, M. and Latour, B. (1992). "Don't Throw the Baby out with the Bath School! A Reply to Collins and Yearley" In A. Pickering (ed.), *Science as Practice and Culture*, Chicago: The University of Chicago Press.

Campbell, J. (1980). "Locke on Qualities." *Canadian Journal of Philosophy X*, 4: 567–585.

Carnap, R. (1928/1967). *The Logical Structure of the World*. Berkeley: University of California Press.

  (1950). *Logical Foundations of Probability*. London: Routledge & Kegan Paul.

  (1957). "Introductory Remarks to the English Translation." In H. Reichenbach *The Philosophy of Space & Time*, pp. v-vii. New York: Dover.

  (1963). "Intellectual Autobiography." In P. Schilpp (ed.), *The Philosophy of Rudolf Carnap*, pp. 3–84. La Salle: Open Court.

  (1966/1995). *An Introduction to the Philosophy of Science*. New York: Dover.

Cartwright, N. (1983). *How the Laws of Physics Lie*. Oxford: Oxford University Press.

Cavell, S. (1979). *The Claim of Reason*. Oxford: Oxford University Press.

Chang, H. (2012a). "Incommensurability: Revisiting the Chemical Revolution." In V. Kindi and T. Arabatzis (eds.), *Kuhn's The Structure of Scientific Revolutions Revisited*, pp. 153–176. New York: Routledge.

  (2012b). *Is Water H_2O?: Evidence, Realism and Pluralism*. Dordrecht: Springer.

  (2018). "Is Pluralism Compatible with Scientific Practice?" In J. Saatsi (ed.), *Routledge Handbook of Scientific Realism*, pp. 176–186. London: Routledge.

Chapman, A. (1984). "Tycho Brahe in China: The Jesuit Mission to Peking and the Iconography of European Instrument Making Processes." *Annals of Science* 41: 417–443.

Chen, X. and Barker, P. (2009). "Process Concepts and Cognitive Obstacles to Change: Perspectives on the History of Science and Science Policy." *Centaurus* 51: 314–320.

Chirimuuta, M. (2015). *Outside Color: Perceptual Science and the Puzzle of Color in Philosophy*. Cambridge, MA: The MIT Press.

Churchland, P. (1979). *Scientific Realism and the Plasticity of Mind*. Cambridge: Cambridge University Press.

(1988). "Perceptual Plasticity and Theoretical Neutrality: A Reply to Jerry Fodor." *Philosophy of Science* 55: 167–187.
Coffa, J. A. (1977). "Carnap's Sprachanschauung Circa 1932." In P. Asquith, and F. Suppe, (eds.), *PSA 1976*, vol. 2. Lansing: PSA.
  (1991). *The Semantic Tradition from Kant to Carnap*. Cambridge: Cambridge University Press.
Cohen, I. B. (1985). *Revolution in Science*. Cambridge, MA: Harvard University Press.
Colyvan, M. (1998). "Can the Eleatic Principle be Justified?" *Canadian Journal of Philosophy* 28, 3: 313–336.
Conant, J. B. (1947/1957). *On Understanding Science: An Historical Approach*. New York: Mentor Books.
  (1943). "Excerpts from Conant Valedictory Address." *Crimson*, January 11. www.thecrimson.com/article/1943/1/11/excerpts-from-conant-valedictory-address-pgentlemen/
  (1946). "Civil Courage." In A. C. Baird (ed.) *Representative American Speeches, 1945–1946*, pp. 223–228. New York: H.W. Wilson.
  (1947). *On Understanding Science*. New Haven: Yale University Press.
  (1957). Foreword. In *The Copernican Revolution: Planetary Astronomy in the Development of Western Thought*, pp. xiii–xviii. Cambridge, MA: Harvard University Press.
  (1970). *My Several Lives: Memoirs of a Social Inventor*. New York: Harper and Row.
Conant, J. (2020). "Reply to Gustafsson: Wittgenstein on the Relation of Sign to Symbol." In S. Miguens (ed.) *The Logical Alien. Conant and His Critics*, pp. 863–947. Cambridge, MA: Harvard University Press.
Conant, J. and Haugeland, J. (2000). *The Road Since "Structure": Philosophical Essays, 1970–1993, with an Autobiographical Interview*. Chicago: University of Chicago Press.
Copernicus, N. (1976). *On the Revolutions of the Heavenly Spheres*, tr. A. M. Duncan. New York: Barnes and Noble.
Crimson, The (1953). "Smith Professor Bares 'Red Cell' Here in 30s," February 26. www.thecrimson.com/article/1953/2/26/smith-professor-bares-red-cell-here/
  (1953a). "Hawkins Confesses Red Affiliations at Hearings," May 11. www.thecrimson.com/article/1953/5/11/hawkins-confesses-red-affiliations-at-hearings/
  (1953b). "Colorado Senate Feuds Defame Three Teachers," May 15. www.thecrimson.com/article/1953/5/15/colorado-senate-feuds-defame-three-teachers/
  (1953c). "Pusey Answers Communism Charge; McCarthy to Cite Furry for Contempt," November 10. www.thecrimson.com/article/1953/11/10/pusey-answers-communism-charge-mccarthy-to/
Crowther, K. and Barker, P. (2013). "Training the Intelligent Eye: Understanding Illustrations in Early Modern Astronomy Texts." *Isis* 104: 429–470.

Curley, E. M. (1972). "Locke, Boyle, and the Distinction between Primary and Secondary Qualities." *Philosophical Review* 81:38–64.
D'Agostino, F. (2005). "Kuhn's Risk-Spreading Argument and the Organization of Scientific Communities." *Episteme* 1, 3: 201–209.
Danielson, D. (2014). Paradise Lost *and the Cosmological Revolution*. Cambridge: Cambridge University Press.
Daston, L. (2016). "History of Science without Structure." In R. Richards and L. Daston (eds.), *Kuhn's 'Structure of Scientific Revolutions' at Fifty: Reflections on a Science Classic*, pp. 115–132. Chicago: University of Chicago Press.
Davidson, D. (1974). "On the Very Idea of a Conceptual Scheme." *Proceedings and Addresses of the American Philosophical Association* 47, 5: 5–20.
Davis Jr., R. (2019). Biographical.NobelPrize.org. Nobel Media AB 2019. Wed. 20 November 2019. www.nobelprize.org/prizes/physics/2002/davis/biographical/
De Langhe, R. (2017). "Towards the Discovery of Scientific Revolutions in Scientometric Data." *Scientometrics* 110:505–519.
Demos, R. (1936). "On Persuasion." *The Journal of Philosophy* 29, 9: 225–232.
Deng, K. H. (2011). "The Cosmology in *Wuweilizhi*." *Journal of Dialectics of Nature* 33, 1: 36–43.
Devlin, W. J. (2015). "An Analysis of Truth in Kuhn's Philosophical Enterprise." In W. J. Devlin, and A. Bokulich (eds.), *Kuhn's Structure of Scientific Revolutions – 50 Years On*, pp. 153–166. Boston Studies in the Philosophy and History of Science, vol. 311: Springer.
Devlin, W. and Bokulich, A. (eds.) (2015). *Kuhn's Structure of Scientific Revolutions – 50 Years On*. Boston Studies in the Philosophy and History of Science, vol. 311: Springer.
DeVries, W. (2005). *Wilfrid Sellars*. London: Routledge.
Dewey, J. (1938). *Logic. The Theory of Inquiry*. New York: Henry Holt and Company.
Dinis, A. (2017). *A Jesuit Against Galileo? The Strange Case of Giovanni Battista Riccioli's Cosmology*. Braga: Axioma.
Duhem, P. (1982/1914). *The Aim and Structure of Physical Theory*. Princeton, NJ: Princeton University Press.
Earley, J. E., (2005). "Why There is No Salt in the Sea." *Foundations of Chemistry* 7: 85–102. https://doi.org/10.1023/B:FOCH.0000042881.05418.15
Ereshefsky, M. (2014). "Species and Taxonomy." In M. Curd and S. Psillos (eds.), *The Routledge Companion to Philosophy of Science*, 2nd ed., pp. 523–530. London: Routledge.
Escobar, L. E. and Craft, M. (2016). "Advances and Limitations of Disease Biogeography Using Ecological Niche Modeling." *Frontiers in Microbiology* 07. 10.3389/fmicb.2016.01174.
Fazlıoğlu, İ. (2007). "Qūshjī: Abū al-Qāsim ʿAlāʾ al-Dīn ʿAlī ibn Muḥammad Qushči-zāde." In T. Hockey et al. (eds.), *The Biographical Encyclopedia of Astronomers, Springer Reference*, pp. 946–948. New York: Springer, 2007.

Felappi, G. (2017). "Susanne Langer and the Woeful World of Facts." *Journal for the History of Analytical Philosophy* 5, 2: 38–50.
Ferrari, M. (2012). "Between Cassirer and Kuhn." *Studies in the History and Philosophy of Science* 43: 18–26.
Feyerabend, P. (1962). "Explanation, Reduction and Empiricism." In H. Feigl and G. Maxwell (eds.), *Scientific Explanation, Space, and Time* (Minnesota Studies in the Philosophy of Science, vol. III), pp. 28–97. Minneapolis: University of Minneapolis Press.
  (1975/1988). *Against Method*, revised ed. London: Verso.
Flexner, A. (1939/2017). *The Usefulness of Useless Knowledge*. Princeton: Princeton University Press.
Fodor, J. (1984). "Observation Reconsidered." *Philosophy of Science* 51: 21–43.
  (1988). "A Reply to Churchland's 'Perceptual Plasticity and Theoretical Neutrality'." *Philosophy of Science* 55: 188–198.
Forrester, J. (2007). "On Kuhn's Case: Psychoanalysis and the Paradigm." *Critical Inquiry* 33 (Summer): 782–819.
Frank, P. (1949). *Modern Science and Its Philosophy*. Cambridge, MA: Harvard University Press.
Friedman, M. (1992). *Kant and the Exact Sciences*. Cambridge: Harvard University Press.
  (2002). "Kant, Kuhn, and the Rationality of Science." In *History of Philosophy of Science*, vol. 9, Vienna Circle Institute Yearbook, pp. 25–41. Dordrecht: Springer.
  (2008). "Ernst Cassirer and Thomas Kuhn." *Philosophical Forum* 39, 2: 239–252.
Galilei, G. (2001). *Dialogue Concerning the Two Chief World Systems: Ptolemaic and Copernican*, tr. S. Drake. New York: Modern Library.
  (2016). *Sidereus Nuncius, or the Sidereal Messenger*, tr. A. Van Helden. Chicago: University of Chicago Press.
Galilei, G. and Scheiner, C. (2010). *On Sunspots*, tr. E. Reeves and A. Van Helden. Chicago: University of Chicago Press.
Galison, P. (1981). "Kuhn and the Quantum Controversy." *British Journal for the Philosophy of Science* 32, 1: 71–85.
  (1990). "Aufbau/Bauhaus: Logical Positivism and Architectural Modernism." *Critical Inquiry* 16, 4: 709–752.
  (1997). *Image and Logic*. Chicago: University of Chicago Press.
  (2016). "Practice All the Way Down." In R. J. Richards and L. Daston (eds.), *Kuhn's 'Structure of Scientific Revolutions' at Fifty: Reflections on a Science Classic*, pp. 42–67. Chicago: University of Chicago Press.
Garfield, E. (1979–1980). "Most-cited Authors in the Arts and Humanities, 1977–1978." *Essays of an Information Scientist* 4: 238–243.
Gattei, S. (2008). *Thomas Kuhn's 'Linguistic Turn' and the Legacy of Logical Empiricism: Incommensurability, Rationality and the Search for Truth*. Aldershot: Ashgate.
Giere, R. (1988). *Explaining Science: A Cognitive Approach*. Chicago: University of Chicago Press.

(2004). "How Models are Used to Represent Reality." *Philosophy of Science* 71: 742–752.

(2006). *Scientific Perspectivism*. Chicago: University of Chicago Press.

Goldstein, B. R. (1967). "The Arabic Version of Ptolemy's Hypotheses." *Transactions of the American Philosophical Society*. New Series, 57, 4: 3–55.

Goldstein, B. and Hon, G. (2007). "Kepler's Move from Orbs to Orbits: Documenting a Revolutionary Scientific Concept." *Perspectives on Science* 3: 74–110.

Gombrich, E. (1950/1995). *The Story of Art*. London: Phaidon Press.

Goodwin, W. (2013). "Structure and Scientific Controversies." *Topoi* 32: 101–110.

(2015). "Revolution and Progress in Medicine." *Theoretical Medicine and Bioethics* 36: 25–39.

Graney, C. M. (2015). *Setting Aside All Authority: Giovanni Battista Riccioli and His Science against Copernicus in the Age of Galileo*. Notre Dame, IN: University of Notre Dame Press.

Grant, E. (1996). *Planets, Stars and Orbs: The Medieval Cosmos, 1200–1687*. Cambridge: Cambridge University Press.

Green, E. (2016). "What are the Most-cited Publications in the Social Sciences (according to Google Scholar)," https://blogs.lse.ac.uk/impactofsocialsciences/2016/05/12/what-are-the-most-cited-publications-in-the-social-sciences-according-to-google-scholar/, (accessed February 24, 2020).

Greif, M. (2015). *The Age of the Crisis of Man*. Princeton: Princeton University Press.

Griffel, F. (2009). *Al-Ghazālī's Philosophical Theology*. New York: Oxford University Press.

(Forthcoming). Ashʿarite Occasionalist Cosmology, al-Ghazālī and the Pursuit of the Natural Sciences in Islamicate Societies. In S. Brentjes (ed.), *The Routledge Handbook of Scientific Practices in Islamicate Societies (8th-19th centuries)*. London: Routledge.

Grube, D. M. (2013) "Interpreting Kuhn's Incommensurability-Thesis: Its Different Meanings and Epistemological Consequences." *Philosophy Study* 3, 5: 377–397.

Hacking, I. (1983). *Representing and Intervening: Introductory Topics in the Philosophy of Natural Science*. Cambridge: Cambridge University Press.

Hadamard, J. (1945). *An Essay on the Psychology of Invention in the Mathematical Field*. Princeton: Princeton University Press.

Hamilton, K. (2001). "Some Philosophical Consequences of Wittgenstein's Aeronautical Research." *Perspectives on Science* 9, 1: 1–37.

Hanson, N. R. (1958). *Patterns of Discovery*. Cambridge: Cambridge University Press.

(n.d.). "Report on Dr. T. Kuhn's *The Structure of Scientific Revolutions*." University of Chicago Press Papers, University of Chicago Special Collections, box 278, folder 4.

Hartner, W. (1969). Nasir al-Din al-Ṭūsī's Lunar Theory. *Physis* 11: 287–304.

(1973). Copernicus, the Man, the Work, and Its History. *Proceedings of the American Philosophical Society* 117: 413–422.
Harvard University (1945). Committee on the Objectives of a General Education in a Free Society. *The Objectives of a General Education in a Free Society*. Cambridge: Harvard University Press.
Hasse, D. N. (2016). *Success and Suppression: Arabic Sciences and Philosophy in the Renaissance*. Cambridge, MA: Harvard University Press.
Heidegger, M. and Gendlin, E. T. (1985). *What is a Thing?* Lanham, MD: University Press of America.
Heilbron, J. L. (1998). "Thomas Samuel Kuhn, 18 July 1922–17 June 1996." *Isis* 89, 3: 505–515.
  (2016). "Where to Start?" In A. Blum, et al. (eds.), *Shifting Paradigms: Thomas S. Kuhn and the History of Science*, pp. 3–13. Berlin: Edition Open Access.
Hempel, C. G. (1980). "Comments on Goodman's *Ways of Worldmaking*." *Synthese* 45: 193–199.
  (2001). *The Philosophy of Carl G. Hempel*, ed. J. Fetzer. New York: Oxford University Press.
Hershberg, J. (1993). *James B. Conant: Harvard to Hiroshima and the Making of the Nuclear Age*. New York: Knopf.
Hesse, M. (1961). *Forces and Fields*. London: Thomas Nelson.
  (1963/1966). *Models and Analogies in Science*. London: Sheed and Ward. 2nd, revised ed. Notre Dame, IN: Notre Dame University Press.
Hirst, R. J. (1967). "Primary and Secondary Qualities." In *The Encyclopedia of Philosophy*, ed. by P. Edwards, pp. 455–457. New York: Macmillan.
Hintikka, J. (1996). "Contemporary Philosophy and the Problem of Truth." *Acta Philosophica Fennica* 6, 1: 23–29.
Hintikka, M. B., and Hintikka, J. (1986). *Investigating Wittgenstein*. Oxford: Basil Blackwell.
Holton, G. (2006). "Philipp Frank at Harvard University." *Synthese* 153, 2: 297–311.
Hoyningen-Huene, P. (1993). *Reconstructing Scientific Revolutions: Thomas S. Kuhn's Philosophy of Science*. Chicago: University of Chicago Press.
  (1994). "Niels Bohr's Argument for the Irreducibility of Biology to Physics." In J. Faye and H. Folse (eds.), *Niels Bohr and Contemporary Philosophy*, pp. 231–255. Dordrecht: Kluwer.
  (2015). "Kuhn's Development before and after *Structure*." In W. J. Devlin and A. Bokulich (eds.) *Kuhn's Structure of Scientific Revolutions: 50 Years On*, pp. 185–195. Boston Studies in the Philosophy and History of Science, vol. 311: Springer.
Hoyningen-Huene, P. and Oberheim, E. (2009). "Reference, Ontological Replacement and Neo-Kantianism: A Reply to Sankey." *Studies In History and Philosophy of Science Part A* 40, 2: 203–209.
Hoyningen-Huene, P., Oberheim, E. and Andersen, H. (1996). "On Incommensurability." *Studies in History and Philosophy of Science* 27, 1: 131–141.

Hufbauer, K. (2012). "From Student of Physics to Historian of Science: T. S. Kuhn's Education and Early Career." *Physical Perspectives* 14: 421–70.
Hutcheson, K. (1987). "Towards a Political Iconology of the Copernican Revolution." In P. Curry (ed.) *Astrology, Science and Society: Historical Essays*, pp. 94–141. Woodbridge, Suffolk: Boydell Press.
Ibn al-Haytham, (1971). *Doubts about Ptolemy* (Shukūk 'alā Baṭlamyūs). ed. A. I. Sabra. [al-Qāhira]: Maṭba'at Dār al-Kutub.
Iliffe. R. (2017). *Priest of Nature: The Religious Worlds of Isaac Newton*. New York: Oxford University Press.
James, W. (1981). *Pragmatism*. Indianapolis: Hackett, (orig. Longmans, Green, 1907).
Jardine, N. (1984). *The Birth of History and Philosophy of Science: Kepler's A Defence of Tycho against Ursus*. Cambridge: Cambridge University Press.
Jardine, N. and Segonds, A. P. (2008). *La guerre des astronomes: la querelle au sujet de l'origine du système géo-héliocentrique à la fin du XVIe siècle*. Paris: Les Belles Lettres.
Jost, W. and Hyde, M. (eds.) (1997). "Introduction. Rhetoric and Hermeneutics." In *Rhetoric and Hermeneutics in Our Time*. pp. 1–44. New Haven: Yale University Press.
Kaempffert, W. (1938). "Toward Bridging the Gaps between the Sciences." *New York Times*, August 7. www.nytimes.com/1938/08/07/archives/toward-bridging-the-gaps-between-the-sciences-an-encyclopedia.html
Kaiser, D. (2016). "Thomas Kuhn and the Psychology of Scientific Revolutions." In R. J. Richards and L. Daston (eds.), *Kuhn's 'Structure of Scientific Revolutions' at Fifty: Reflections on a Science Classic*, pp. 71–95. Chicago: University of Chicago Press.
Kant, I. (1781&1787/1998). *Critique of Pure Reason*, ed., tr. P. Guyer and A. Wood. Cambridge: Cambridge University Press.
  (1783/2004) *Prolegomena to Any Future Metaphysics That Will be Able to Come Forward as Science: With Selections from the Critique of Pure Reason*, ed., tr. G. C. Hatfield. Rev. ed. Cambridge: Cambridge University Press.
Kaufman, J. C. and Sternberg, R. J. (2010). *The Cambridge Handbook of Creativity*. Cambridge: Cambridge University Press.
Kaufmann, T. D. (1993). *The Mastery of Nature: Aspects of Art, Science, and Humanism in the Renaissance*. Princeton: Princeton University Press.
Kaulbach, F. (1973). "Die Copernicanische Denkfigur bei Kant." *Kant-Studien* 64, 1: 30–48.
Keating, L. (1993). "Un-Locke-ing Boyle: Boyle on Primary and Secondary Qualities." *History of Philosophy Quarterly* 10: 305–323.
Kennedy, E. S. (1966). "Late Medieval Planetary Theory." *Isis* 57: 365–378. Reprinted in E. S. Kennedy, et al. (1983). *Studies in the Islamic Exact Sciences*, pp. 84–97. Beirut: American University of Beirut.
  (1971). "Planetary Theory in the Medieval Near East and Its Transmission to Europe." In E. Cerullo et al. (eds.), *Orient e Occidente nel Medioevo: filosophia e scienze*, pp. 98–107. Rome, Academia Nazionale dei Lincei.

Kennedy E. S. and Roberts, V. (1959). "The Planetary Theory of Ibnal-Shāṭir." *Isis* 50: 227–235.

Kepler, J. (2015). *Astronomia Nova*. Revised ed. tr. W. H. Donahue. Santa Fe, NM: Green Lion Press.

Kindi, V. (2009). "Second Thoughts on Wittgenstein's Secondary Sense." In V. A. Munz, K. Puhl, and J. Wang (eds.), *Language and World*, pp. 202–204. 32nd International Wittgenstein Symposium, Kirchberg am Wechsel.

(2010). "Novelty and Revolution in Art and Science: The Influence of Kuhn on Cavell." *Perspectives on Science* 18, 3: 284–310.

(2012). "Kuhn's Paradigms." In V. Kindi and T. Arabatzis (eds.), *Kuhn's The Structure of Scientific Revolutions Revisited*, pp. 91–111. New York: Routledge.

(2016). "Wittgenstein and Philosophy of Science." In H. J. Glock and J. Hyman (eds.), *A Companion to Wittgenstein*, pp. 587–602. New York: Wiley Blackwell.

Kindi, V. and Arabatzis, T. (eds.) (2012). *Kuhn's The Structure of Scientific Revolutions Revisited*. New York: Routledge.

King, D. A. (1975). "Ibn al-Shāṭir." In C. C. Gillispie (ed.), *Dictionary of Scientific Biography*, 18, 7: 357–364.

(2007). "Ibn al-Shāṭir: 'Alā' al-Dīn 'Alī ibn Ibrāhīm." In T. Hockey, et al. (eds.), *The Biographical Encyclopedia of Astronomers*, pp. 569–570. Springer Reference. New York: Springer.

King, D. A., Samsó, J. and Goldstein, B. R. (2001). "Astronomical Handbooks and Tables from the Islamic World (750–1900): An Interim Report." *Suhayl* 2: 9–105.

Kitcher, P. (1993). *The Advancement of Science: Science without Legend, Objectivity without Illusions*. Oxford: Oxford University Press.

Klein, M. (1970). *Paul Ehrenfest, vol. I*. Amsterdam: North-Holland.

Klein, M., Shimony, A., and Pinch, T. (1979). "Paradigm Lost?" *Isis* 70: 429–440.

Koestler, A. (1981/1985). "The Three Domains of Creativity." In D. Duttoon and M. Krausz (eds.), *The Concept of Creativity in Science and Art*, pp. 1–17. Boston: Martinus Nijhoff Publishers.

Koyré, A. (1939). *Études galiléennes*. Paris: Hermann.

Kragh, H. (2000). "Conceptual Changes in Chemistry: The Notion of Chemical Elements, ca. 1900–1925." *Studies in History and Philosophy of Physics* 31: 435–450.

Kuhn, T. S. (1941). "The War and My Crisis," TSK Archives–MC240, box 1, folder 3.

(1943) "Phi Beta Kappa Address," TSK Archives–MC240, box 1, folder 3.

(1945). "Subjective View. Thomas S. Kuhn, on Behalf of the Recent Students, Reflects Undergraduate Attitude." *Harvard Alumni Bulletin*, September 22, 29–30.

(1951a). Letter to Ralph Lowell, February 20, TSK Archives–MC240, box 3, folder 10.

(1951b). Letter to Dean Owen, January 6, TSK Archives–MC240, box 3, folder 10.

(1953). Letter to Charles Morris, July 3, TSK Archives—MC240, box 25, folder 53.
(1955). "Can the Layman Know Science? State Teachers College—Bridgewater, Massachusetts 12/13/55," TSK Archives—MC240, box 3, folder 33.
(1957/2003). *The Copernican Revolution: Planetary Astronomy in the Development of Western Thought*. Cambridge, MA: Harvard University Press.
(1959/1977). "The Essential Tension: Tradition and Innovation in Scientific Research." In *The Essential Tension: Selected Studies in Scientific Tradition and Change*, pp. 225–239. Chicago: University of Chicago Press.
(1962a). *The Structure of Scientific Revolutions*, Chicago: University of Chicago Press.
(1962b) Letter from Thomas S. Kuhn to Edwin B. Boring, November 29, TSK Archives— MC 240, box 4, folder 7.
(1962/1970). *The Structure of Scientific Revolutions*. 2nd ed. Chicago: University of Chicago Press. The second, 1970, edition includes "Postscript-1969."
(1962/1977). "The Historical Structure of Scientific Discovery." In *The Essential Tension: Selected Studies in Scientific Tradition and Change*, pp. 165–177. Chicago: University of Chicago Press.
(1962/1996). *The Structure of Scientific Revolutions*, 3rd ed. Chicago: University of Chicago Press.
(1962/2012). *The Structure of Scientific Revolutions*, 4th ed. *50th anniversary edition. With an Introductory Essay by Ian Hacking*. Chicago, University of Chicago Press.
(1963). "The Function of Dogma in Scientific Research." In A. C. Crombie (ed.), *Scientific Change: Historical Studies in the Intellectual, Social and Technical Conditions for Scientific Discovery and Technical Innovation, from Antiquity to the Present*, pp. 347–369. London: Heinemann Educational Books.
(1964/1977). "A Function for Thought Experiments." In *The Essential Tension: Selected Studies in Scientific Tradition and Change*, pp. 240–265. Chicago: University of Chicago Press.
(1967). "Paradigms and Theories in Scientific Research." TSK Archives—MC 240, box 3, folder 14. Cited from Marcum 2012a.
(1969/1974). "Second Thoughts on Paradigms – 1969 lecture" In F. Suppe (ed.), *The Structure of Scientific Theories*, pp. 459–482. Urbana, IL: University of Illinois Press.
(1969/1977). "Comment on the Relations of Science and Art." In *The Essential Tension: Selected Studies in Scientific Tradition and Change*, pp. 340–351. Chicago: University of Chicago Press.
(1969/2012). "Postscript – 1969." In Fourth edition, 50th anniversary edition. With an Introductory Essay by Ian Hacking, pp. 173–208.
(1970/2000) "Reflections on My Critics." In J. Conant and J. Haugeland (eds.), *The Road since Structure: Philosophical Essays, 1970–1993, with an Autobiographical Interview*, pp. 123–175, Chicago: Chicago University Press.

(1970/1977). "Logic of Discovery or Psychology of Research?" In *The Essential Tension: Selected Studies in Scientific Tradition and Change*, pp. 266–292. Chicago: University of Chicago Press.

(1971/1977). "Concepts of Cause in the Development of Physics." In *The Essential Tension: Selected Studies in Scientific Tradition and Change*, pp. 21–30. Chicago: University of Chicago Press.

(1973). Letter to Kenneth Pietrzak, April 17, TSK Archives–MC240, box 10.

(1973/1977). "Objectivity, Value Judgment, and Theory Choice." In *The Essential Tension: Selected Studies in Scientific Tradition and Change*, pp. 320–339. Chicago: University of Chicago Press.

(1974/1977). "Second Thoughts on Paradigms." In *The Essential Tension: Selected Studies in Scientific Tradition and Change*, pp. 293–319. Chicago: University of Chicago Press.

(1976/1977). "The Relation between the History and the Philosophy of Science." In *The Essential Tension: Selected Studies in Scientific Tradition and Change*, pp. 3–20. Chicago: University of Chicago Press.

(1976/2000). "Theory-Change as Structure-Change: Comments on the Sneed Formalism." In J. Conant and J. Haugeland (eds.), *The Road since Structure: Philosophical Essays, 1970–1993, with an Autobiographical Interview*, pp. 176–195, Chicago: Chicago University Press.

(1977a). *The Essential Tension: Selected Studies in Scientific Tradition and Change*. Chicago: The University of Chicago Press.

(1977b). "Preface." In *The Essential Tension: Selected Studies in Scientific Tradition and Change*, pp. ix–xxiii. Chicago: University of Chicago Press.

(1977c). "Objectivity, Value Judgment, and Theory Choice." In *The Essential Tension: Selected Studies in Scientific Tradition and Change*, pp. 320–339. Chicago: University of Chicago Press.

(1978). *Black-Body Theory and the Quantum Discontinuity, 1894–1912*. Oxford: Oxford University Press.

(1979). "Metaphor in Science." In A. Ortony (ed.), *Metaphor and Thought*, pp. 409–419. Cambridge: Cambridge University Press.

(1979/2000). "Metaphor in Science." In J. Conant and J. Haugeland (eds.), *The Road since Structure: Philosophical Essays, 1970–1993, with an Autobiographical Interview*, pp. 196–207, Chicago: Chicago University Press.

(1983/2000). "Commensurability, Comparability, Communicability." In J. Conant and J. Haugeland (eds.), *The Road since Structure: Philosophical Essays, 1970–1993, with an Autobiographical Interview*, pp. 33–57, Chicago: Chicago University Press.

(1984). "Lecture IV – Conveying the Past to the Present." Lectures/Meetings: Thalheimer Lectures, "Scientific Development and Lexical Change," TSK Archives—MC 240, box 23.

(1984/2017). *Desarrollo científico y cambio de léxico. Conferencias Thalheimer*, ed. P. Melogno. Montevideo: ANII/UdelaR/SADAF.

(1987/2000). "What are Scientific Revolutions?" In J. Conant and J. Haugeland (eds.), *The Road since Structure: Philosophical Essays, 1970–1993, with an Autobiographical Interview*, pp. 13–32, Chicago: Chicago University Press.

(1989/2000). "Possible Worlds in History of Science." In J. Conant and J. Haugeland (eds.), *The Road since Structure: Philosophical Essays, 1970–1993, with an Autobiographical Interview*, pp. 58–89, Chicago: Chicago University Press.

(1990). "The Road since Structure." In *PSA: Proceedings of the Biennial Meeting of the Philosophy of Science Association*, vol. 1990, Volume Two: Symposia and Invited Papers, pp. 3–13.

(1991/2000). "The Road since *Structure*." In J. Conant and J. Haugeland (eds.), *The Road since Structure: Philosophical Essays, 1970–1993, with an Autobiographical Interview*, pp. 90–104, Chicago: Chicago University Press.

(1992/2000). "The Trouble with the Historical Philosophy of Science." In J. Conant and J. Haugeland (eds.), *The Road since Structure: Philosophical Essays, 1970–1993, with an Autobiographical Interview*, pp. 105–120, Chicago: Chicago University Press.

(1993/2000). "Afterwords." In J. Conant and J. Haugeland (eds.), *The Road since Structure: Philosophical Essays, 1970–1993, with an Autobiographical Interview*, pp. 224–252, Chicago: Chicago University Press.

(2000a). *The Road since Structure: Philosophical Essays, 1970–1993, with an Autobiographical Interview*, eds. J. Conant and J. Haugeland. Chicago: Chicago University Press.

(2000b). "A Discussion with Thomas Kuhn" (with Aristides Baltas, Kostas Gavroglu, Vasso Kindi). In J. Conant and J. Haugeland (eds.), *The Road since Structure: Philosophical Essays, 1970–1993, with an Autobiographical Interview*, pp. 253–323, Chicago: Chicago University Press.

(n.d.) Untitled document ("Dear Professor Frank"), TSK Archives—MC240, box 25, folder 53.

(n.d.) Kuhn M1: SSR – Chapter 1 – "What are Scientific Revolutions?" TSK Archives—MC240, box 4, folder 3.

(n.d.) Kuhn M2: SSR – Chapter 1 – "Discoveries as Revolutionary." TSK Archives—MC240, box 4, folder 3.

(n.d.) *The Plurality of Worlds: An Evolutionary Theory of Scientific Development*. Unpublished manuscript.

Kusch, M. (1989). *Language as Calculus vs. Language as Universal Medium*. Dordrecht Boston: Kluwer Academic Publishers.

Kusuba, T. and Pingree, D. (2002). *Arabic Astronomy in Sanskrit*. Leiden: Brill.

Kuukkanen, Jouni-Matti. (2007). "Kuhn, the Correspondence Theory of Truth and Coherentist Epistemology." *Studies in History and Philosophy of Science* 38: 555–566.

(2012). "The Concept of Evolution in Kuhn's Philosophy." In V. Kindi and T. Arabatzis (eds.), *Kuhn's The Structure of Scientific Revolutions Revisited*, pp. 134–153. New York: Routledge.

Lakatos, I. (1971). "History of Science and Its Rational Reconstructions." In R. Buck and R. S. Cohen (eds.), *PSA: Proceedings of the Biennial Meeting of the Philosophy of Science Association 1970*, Dordrecht: Reidel. Reprinted in I. Lakatos (1978) *The Methodology of Scientific Research Programmes: Philosophical Papers*, vol. 1, pp. 102–138. Cambridge: Cambridge University Press.

(1978). *The Methodology of Scientific Research Programmes: Philosophical Papers, vol. 1*. Cambridge: Cambridge University Press.

Langer, S. (1942/1951). *Philosophy in a New Key*, 2nd ed. New York: Mentor Books.

Laudan, L. (1984). *Science and Values: The Aims of Science and their Role in Scientific Debate*. Berkeley and Los Angeles: University of California Press.

Launert, D. (2009). Le système du monde de Nicolas Raimar Ursus comparé à ceux de Brahe et Roeslin. In M. Á. Granada and É. Mehl (eds.), *Nouveau Ciel, Nouvelle Terre: La révolution copernicienne dans l'Allemagne de la Réforme (1530–1630)*, pp. 155–178. Paris: Les Belles Lettres.

Lee, M.-K. (2011). "The Distinction between Primary and Secondary Qualities in Ancient Greek Philosophy." In L. Nola (ed.) *Primary and Secondary Qualities: The Historical and Ongoing Debate*, pp. 15–40. Nola. Oxford: Oxford University Press.

Lerner, M.-P. (1995). "L'entrée de Tycho Brahe chez les jésuites ou le chant du cygne de Clavius." In L. Giard (ed.), *Les jésuites à la Renaissance: Système educatif et production du savoir*, pp. 145–185. Paris: PUF.

(2008). *Le monde des sphères*. 2 vols. 2nd ed. Paris: Les Belles Lettres.

Lipton, P. (2003). "Kant on Wheels." *Social Epistemology* 17, 2–3: 215–219.

MacBride, F. (2019). "Truthmakers." In E. N. Zalta (ed.), *The Stanford Encyclopedia of Philosophy*. (Spring 2019 edition), https://plato.stanford.edu/archives/spr2019/entries/truthmakers/

Macintosh, J. J. (1976). "Primary and Secondary Qualities." *Studia Leibnitiana* 8:88–104.

MacIntyre, A. (1977). "Epistemological Crises, Dramatic Narrative, and the Philosophy of Science." *The Monist* 60: 453–471.

Marcum, J. (2012a). "From Paradigm to Disciplinary Matrix and Exemplar." In V. Kindi and T. Arabatzis (eds.), *Kuhn's The Structure of Scientific Revolutions Revisited*, pp. 41–63. New York: Routledge.

(2012b). "Wither Thomas Kuhn's Historical Philosophy of Science? An Evolutionary Turn." *Athens: Atiner's Conference Paper Series*, No. PHI2012-0088.

(2015a). "The Evolving Notion and Role of Kuhn's Incommensurability Thesis." In W. J. Devlin, and A. Bokulich (eds.), *Kuhn's Structure of Scientific Revolutions – 50 Years On*, pp. 115–134. Boston Studies in the Philosophy and History of Science, vol. 311: Springer.

(2015b). *Thomas Kuhn's Revolutions: A Historical and an Evolutionary Philosophy of Science?* London: Bloomsbury.

Margolis, H. (1987). *Patterns, Thinking, and Cognition*. Chicago: University of Chicago Press.
Martinez, J. A. (1974). "Galileo on Primary and Secondary Qualities." *Journal for the History of the Behavioural Sciences* 10: 160–169.
Marx, K. (1904). *A Contribution to the Critique of Political Economy*, tr. N. I. Stone. International Publishers.
Masterman, M. (1970). "The Nature of a Paradigm." In I. Lakatos and A. Musgrave (eds.) *Criticism and the Growth of Knowledge*, pp. 59–89. Cambridge: Cambridge University Press.
Mayoral de Lucas, J. V. (2009). "Intensions, Belief and Science: Kuhn's Early Philosophical Outlook (1940–1945)." *Studies in History and Philosophy of Science* 40: 175–184.
  (2017). *Thomas S. Kuhn: La Búsqueda de la Estructura*. Zaragoza, Spain: Prensas de la Universidad de Zaragoza.
McCann, E. (2011). "Locke's Distinction Between Primary Primary Qualities and Secondary Primary Qualities." In L. Nola (ed.) *Primary and Secondary Qualities: The Historical and Ongoing Debate*, pp. 158–189. Oxford: Oxford University Press.
McGuinness, B. (ed.) (1987). *Unified Science*. Dordrecht: D. Reidel.
McMullin, E. (1976). "The Fertility of Theory and the Unit for Appraisal in Science." In R. S. Cohen, P. K. Feyerabend, and M. W. Wartofsky (eds.), *Essays in Memory of Imre Lakatos*, pp. 395–432. Dordrecht: Reidel.
Melogno, P. (2019). "The Discovery-Justification Distinction and the New Historiography of Science: On Thomas Kuhn's Thalheimer Lectures." *HOPOS* 9, 1: 152–178.
Menand, L. (2001). *The Metaphysical Club*. New York: Farrar, Straus and Giroux.
Mendeleev, D. I. (1891). *Principles of Chemistry*, vol. 1, 1st English ed., tr. G. Kamensky. London: Longmans, Green.
Mill, J. S. (1859/1956). *On Liberty*. New York: The Liberal Arts Press.
Mladenović, B. (2007). "Muckraking in History." *Perspectives on Science* 15, 3: 261–294.
Moore, G. E. (1925). "A Defence of Common Sense." In J. H. Muirhead (ed.) *Contemporary British Philosophy* (2nd series).www.ditext.com/moore/common-sense.html
Moran, B. T. (1981). "German Prince-Practitioners: Aspects in the Development of Courtly Science, Technology, and Procedures in the Renaissance." *Technology and Culture* 22, 2: 253–274.
Mormann, T. (2012). "A Place for Pragmatism in the Dynamics of Reason?" *Studies in the History and Philosophy of Science* 43: 27–37.
  (2017). "Philipp Frank's Austro-American Logical Empiricism." *HOPOS* 7, 1: 56–87.
Morris, C. (1963). "Pragmatism and Logical Empiricism." In P. Schilpp (ed.), *The Philosophy of Rudolf Carnap*, pp. 87–98. La Salle: Open Court.
Morris, E. (2018). *The Ashtray: (Or the Man Who Denied Reality)*. Chicago: University of Chicago Press.

Morrison R. G. (2014). "A Scholarly Intermediary Between the Ottoman Empire and Renaissance Europe." *Isis* 105: 32–57.

Morrison, R. G. (2016). *Joseph ibn Nahmias' The Light of the World: Astronomy in al-Andalus*. Berkeley: University of California Press.

Nagel, E. (1961). *The Structure of Science: Problems in the Logic of Scientific Explanation*. New York: Harcourt, Brace and World.

Nash, L. (1957). "The Atomic-Molecular Theory." In J. Conant (ed.), *Harvard Case Histories in Experimental Science*, vol. 1, pp. 215–321. Cambridge, MA: Harvard University Press.

Nersessian, N. (2008). *Creating Scientific Concepts*. Cambridge, MA: MIT Press.

Neurath, O. (1932/1987). "Unified Science and Psychology." In B. McGuinness (ed.), *Unified Science*, pp. 1–23. Dordrecht: D. Reidel.

　(1937/1987). "The New Encyclopedia." In B. McGuinness (ed.), *Unified Science*, 132–141. Dordrecht: D. Reidel.

Neurath et al. (1929/1973). "The Scientific Conception of the World: The Vienna Circle." In *Empiricism and Sociology*, pp. 299–318. Dordrecht: D. Reidel.

New York Times (1945). "Truman Aid Asked for Magnuson Bill," November 27.

　(1953). "M'Carthy Charges 'Mess' at Harvard," November 6.

Nickles, T. (1973). "Two Concepts of Intertheoretic Reduction." *Journal of Philosophy* 70: 181–201.

　(1976). "Theory Generalization, Problem Reduction, and the Unity of Science." In A. Michalos and R. S. Cohen (eds.) *PSA 1974*, pp. 31–74. Dordrecht: Reidel.

　(1980). "Can Scientific Constraints Be Violated Rationally?" In T. Nickles (ed.), *Scientific Discovery, Logic, and Rationality*, pp. 285–315. Dordrecht: Reidel.

　(1981). "What Is a Problem That We May Solve It?" *Synthese* 47: 85–118.

　(1994). "Enlightenment versus Romantic Models of Creativity in Science—and Beyond." *Creativity Research Journal* 7, 3–4: 277–314.

　(2000). "Kuhnian Puzzle Solving and Schema Theory." *Philosophy of Science* 67: S242–255.

　(2003). "Normal Science: From Logic to Case-Based and Model-Based Reasoning." In T. Nickles (ed.) *Thomas Kuhn*, pp. 142–177. Cambridge: Cambridge University Press.

　(2005). "Problem Reduction: Some Thoughts." In R. Festa, A. Aliseda, and J. Peijnenburg (eds.), *Cognitive Structures in Scientific Inquiry: Essays in Debate with Theo Kuipers*, vol. 2, pp. 107–133. Amsterdam: Rodopi.

　(2006). "Heuristic Appraisal: Context of Discovery or Justification?" In J. Schickore and F. Steinle (eds.), *Revisiting Discovery and Justification: Historical and Philosophical Perspectives on the Context Distinction*, pp. 159–182. Dordrecht: Springer.

　(2012). "Some Puzzles about Kuhn's Exemplars." In V. Kindi and T. Arabatzis (eds.), *Kuhn's The Structure of Scientific Revolutions Revisited*, pp. 112–133. New York: Routledge.

(2017). "Scientific Revolutions." In E. N. Zalta (ed.), *The Stanford Encyclopedia of Philosophy*. First published 5 March 2009. Heavily revised version, 28 November 2017. https://plato.stanford.edu/entries/scientific-revolutions/
Norman, D. (1990). *The Design of Everyday Things*. New York: Doubleday.
(1993). *Things that Make Us Smart*. New York: Basic Books.
Nye, M. J. (1989). "Chemical Explanations and Physical Dynamics: Two Research Schools at the First Solvay Conferences, 1922–1928." *Annals of Science* 46: 461–480.
(2012). "Thomas Kuhn, Case Histories, and Revolutions." *Historical Studies in the Natural Sciences* 42, 5: 557–561.
Oberheim, Eric (2016). "Rediscovering Einstein's Legacy: How Einstein Anticipates Kuhn and Feyerabend on the Nature of Science." *Studies in History and Philosophy of Science Part A* 57: 17–26.
Oberheim, E. and Hoyningen-Huene, P. (2018). "The Incommensurability of Scientific Theories." In E. N. Zalta (ed.), *The Stanford Encyclopedia of Philosophy*. https://plato.stanford.edu/archives/fall2018/entries/incommensurability/
Omodeo, P. D. (2014). *Copernicus in the Cultural Debates of the Renaissance: Reception, Legacy, Transformation*. Leiden: Brill.
Palmer, D. (1976): "Boyle's Corpuscular Hypothesis and Locke's Primary-Secondary Quality Distinction." *Philosophical Studies* 29: 181–189.
Paneth, F. A. (1962). "The Epistemological Status of the Chemical Concept of Element." *British Journal for the Philosophy of Science* 13: 1–14 and 144–160. (Reprinted in *Foundations of Chemistry* 5 (2003): 113–145.)
Patton, L. (2004). *Hermann Cohen's History and Philosophy of Science*. Dissertation, McGill University.
(2017). "Kuhn, Pedagogy, and Practice." In M. Mizrahi (ed.), *The Kuhnian Image of Science: Time for a Decisive Transformation?* pp. 113–130. Lanham, MD: Rowman & Littlefield.
Pearson, K. (1892). *The Grammar of Science*. London: Walter Scott.
Pedersen, O. (1974). *A Survey of the Almagest*. Odense: Odense Universitetsforlag.
(1978). "The Decline and Fall of the *Theorica Planetarum*." In E. Rosen et al. (eds.), *Science and History: Studies in Honor of Edward Rosen*, pp. 157–185. Warsaw: Ossolineum, Polish Academy of Sciences Press.
(1981). "The Origins of the *Theorica Planetarum*." *Journal for the History of Astronomy* 12: 113–123.
(1985). "In Quest of Sacrobosco." *Journal for the History of Astronomy* 16: 175–220.
Piaget, J. (1928). *Judgment and Reasoning in the Child*, tr. M. Warden. New York: Harcourt, Brace.
Pietrzak, K. (1973). Letter to Thomas Kuhn, April 6, TSK Archives–MC240, box 10.
Pihlström, S. and Siitonen, A. (2005). "The Transcendental Method and (Post-)Empiricist Philosophy of Science." *Journal for General Philosophy of Science* 36, 1: 81–106.

Pingree, D. (1999). An Astronomer's Progress. *Proceedings of the American Philosophical Society* 143, 1: 73–85.

Pinto de Oliveira, J. C. (2007). "Carnap, Kuhn, and Revisionism: On the Publication of Structure in Encyclopedia." *Journal for General Philosophy of Science* 38: 147–157.

(2012). "Kuhn and the Genesis of the 'New Historiography of Science.'" *Studies in History and Philosophy of Science* 43: 115–121.

(2014). "History of Science and History of Art: An Introduction to Kuhn's Theory." http://philsci-archive.pitt.edu/11231

(2015). "Carnap, Kuhn, and the History of Science: A Reply to Thomas Uebel." *Journal for General Philosophy of Science* 46: 215–223.

(2017). "Thomas Kuhn, the Image of Science and the Image of Art: The First Manuscript of *Structure*." *Perspectives on Science* 25: 746–765.

(2020). "Kuhn, Condorcet, and Comte: On the Justification of the 'Old' Historiography of Science." *Perspectives on Science* 28: 375–397.

Pinto de Oliveira, J. C. and Oliveira, A. J. (2018). "Kuhn, Sarton, and the History of Science." In R: Pisano, et al. (eds.), *Hypotheses and Perspectives within History and Philosophy of Science. Homage to Alexandre Koyré 1964–2014*, pp. 277–293. Cham: Springer.

Polanyi, M. (1946). *Science, Faith and Society*, Oxford: Oxford University Press.

(1958). *Personal Knowledge*. Chicago: University of Chicago Press.

(1966). *The Tacit Dimension*. Garden City, NY: Doubleday.

Politi, V. (2018). "Scientific Revolutions, Specialization and the Discovery of the Structure of DNA: Toward a New Picture of the Development of the Sciences." *Synthese* 195: 2267–2293.

Popper, K. (1959). *The Logic of Scientific Discovery*. London: Hutchinson. Translation of expanded version of Logic der Forschung, 1934.

(1959/1968). *The Logic of Scientific Discovery*. New York: Harper & Row. (Originally published in German in 1934.)

(1953/1969). "Science: Conjectures and Refutations." In: *Conjectures and Refutations*, pp 43–86. London: Routledge.

(1963). *Conjectures and Refutations: The Growth of Scientific Knowledge*. London: Routledge.

(1970). "Normal Science and its Dangers." In I. Lakatos and A. Musgrave (eds.), *Criticism and the Growth of Knowledge: Proceedings of the International Colloquium in the Philosophy of Science 1965, Volume 4*, pp. 51–58. Cambridge: Cambridge University Press.

Popper, K. et al. (1969/1976). *The Positivist Dispute in German Sociology*. London: Heinemann.

Post, H. (1971). "Correspondence, Invariance and Heuristics: In Praise of Conservative Induction." *Studies in History and Philosophy of Science* 2: 213–255.

Quine, W. V. O. (1964). *From a Logical Point of View*. Cambridge, MA: Harvard University Press.

Raftopoulos, A. (2009). *Cognition and Perception. How Do Psychology and Neural Science Inform Philosophy?* Cambridge, MA: MIT Press.

Ragep, F. J. (2001). "Tusi and Copernicus: The Earth's Motion in Context." *Science in Context* 14, 1/2: 145–163. DOI: 10.1017/0269889701000060

(2007). "Shīrāzī: Quṭb al-Dīn Maḥmūd ibn Masʿūd Muṣliḥ al-Shīrāzī." In T. Hockey et al. (eds.), *The Biographical Encyclopedia of Astronomers.* Springer Reference, pp. 1054–1055. New York: Springer, 2007.

(2008). "Hay'a." In H. Selin (ed.) *Encyclopaedia of the History of Science, Technology, and Medicine in Non-Western Cultures.* Dordrecht: Springer.

Reichenbach, H. (1928/1957) *The Philosophy of Space & Time.* With introductory remarks by Rudolf Carnap. New York: Dover.

(1931). "Aims and Methods of Modern Philosophy of Nature." In *Modern Philosophy of Science*, pp. 79–108. London: Routledge.

(1948) "Rationalism and Empiricism." In *Modern Philosophy of Science*, pp. 135–150. London: Routledge.

(1951). *The Rise of Scientific Philosophy.* Berkeley: University of California Press.

(1959) *Modern Philosophy of Science: Selected Essays by Hans Reichenbach.* London: Routledge.

Reisch, G. A. (1994). "Planning Science: Otto Neurath and the *International Encyclopedia of Unified Science*." *British Journal for the History of Science* 27: 153–175.

(2012). "The Paranoid Style in American History of Science." *Theoria* 27/3, 75: 323–342.

(2016). "Aristotle in the Cold War: On the Origins of Thomas Kuhn's *The Structure of Scientific Revolutions*." In R. J. Richards and L. Daston, (eds.), *Kuhn's 'Structure of Scientific Revolutions' at Fifty: Reflections on a Science Classic*, pp. 12–29. Chicago: University of Chicago Press.

(2019a). *The Politics of Paradigms: Thomas S. Kuhn, James B. Conant, and the Cold War "Struggle for Men's Minds."* Albany: SUNY Press.

(2019b). "What a Difference a Decade Makes: The Planning Debates and the Fate of the Unity of Science Movement." In J. Cat and A. T. Tuboly (eds.), *Neurath Reconsidered*, pp. 385–411. Cham: Springer.

Renzi, B. G. (2009). "Kuhn's Evolutionary Epistemology and Its Being Undermined by Inadequate Biological Concepts." *Philosophy of Science* 76: 143–159.

Rescher, N. (1982). *The Coherence Theory of Truth.* Washington: University Press of America.

Reydon, T. A. C. and Hoyningen-Huene, P. (2010). "Discussion: Kuhn's Evolutionary Analogy in *The Structure of Scientific Revolutions* and 'The Road since Structure.'" *Philosophy of Science* 77: 468–476.

Rheticus, G. (2004). *Narratio prima*, tr. E. Rosen. In E. Rosen, *Three Copernican Treatises.* New York: Dover.

Richards, R. and Daston L. (eds.) (2016). *Kuhn's 'Structure of Scientific Revolutions' at Fifty: Reflections on a Science Classic.* Chicago: University of Chicago Press.

Richardson, A. W. (2002). "Narrating the History of Reason Itself." *Perspectives on Science* 10 (3): 253–274.
  (2003). "Conceiving, Experiencing, and Conceiving Experiencing." *Topoi* 22: 55–67.
  (2008). "Scientific Philosophy as a Topic for History of Science." *Isis* 99, 1: 88–96.
  (2012). "The Structure of Philosophical History." In V. Kindi and T. Arabatzis (eds.), *Kuhn's The Structure of Scientific Revolutions Revisited*, pp. 231–250. New York: Routledge.
  (2015). "From Troubled Marriage to Uneasy Colocation: Thomas Kuhn, Epistemological Revolutions, Romantic Narratives, and History and Philosophy of Science." In W. J. Devlin, and A. Bokulich (eds.), *Kuhn's Structure of Scientific Revolutions – 50 Years On*, pp. 39–50. Boston Studies in the Philosophy and History of Science, vol. 311: Springer.
Richardson, A. and Uebel, T. (eds.). (2007) *The Cambridge Companion to Logical Empiricism*. Cambridge: Cambridge University Press.
Roberts, V. (1957). "The Solar and Lunar Theory of Ibn ash-Shatir." *Isis*, 48: 428–432.
Robinson, K. (2010). "Changing Paradigms." Animated RSA Talk, available at www.youtube.com/watch?v=BHMUXFdBzik, (accessed October 5, 2019).
Roeslin, H. (2000). *De opere Dei creationis seu de mundo hypotheses*, ed. M. Granada. Lecce: Conte.
Roller, D. and Roller, D. H. D. (1957). "The Development of the Concept of Electric Charge: Electricity from the Greeks to Coulomb." In J. Conant (ed.), *Harvard Case Histories in Experimental Science*, vol. 2, pp. 541–639. Cambridge, MA: Harvard University Press.
Rorty, R. (1982). *Consequences of Pragmatism*. Minneapolis, University of Minnesota.
Rosefeldt, T. (2007). "Dinge an sich und sekundäre Qualitäten." In J. Stolzenberg (ed.), *Kant in der Gegenwart*, pp. 167–209. Berlin: de Gruyter.
Rothman, A. (2017). *The Pursuit of Harmony: Kepler on Cosmos, Confession, and Community*. Chicago: University of Chicago Press.
Rouse, J. (1996). *Engaging Science: How to Understand Its Practices Scientifically*. Ithaca: Cornell University Press.
  (1998). "Kuhn and Scientific Practices." *Division I Faculty Publications*, Wesleyan University. Paper 17. http://wesscholar.wesleyan.edu/div1fac pubs/17
  (2013). "Recovering Thomas Kuhn." *Topoi* 32, 1: 59–64.
Rowbottom, D. P. (2011). "What's at the Bottom of Scientific Realism?" *Studies in History and Philosophy of Science Part A* 42, 4: 625–628.
Rueger, A. (1996). "Risk and Diversification in Theory Choice." *Synthese* 109: 263–280.
Runco, M. A. (2010). "Divergent Thinking, Creativity, and Ideation." In J. C. Kaufman and R. J. Sternberg (eds.) *The Cambridge Handbook of Creativity*, pp. 413–446. Cambridge: Cambridge University Press.

Runco, M. A. and Albert, R. S. (2010). "Creativity Research: A Historical View." In J. C. Kaufman and R. J. Sternberg (eds.) *The Cambridge Handbook of Creativity*, pp. 3–19. Cambridge: Cambridge University Press.

Ruphy, S. (2011). "From Hacking's Plurality of Styles of Scientific Reasoning to 'Foliated' Pluralism: A Philosophically Robust Form of Ontologico-Methodological Pluralism." *Philosophy of Science* 78, 5: 1212–1222.

(2013/2016). *Scientific Pluralism Reconsidered: A New Approach to the (Dis)Unity of Science*. Pittsburgh: University of Pittsburgh Press.

Saliba, G. (1987). "The Role of Maragha in the Development of Islamic Astronomy: A Scientific Revolution before the Renaissance." *Revue de Synthese* 108: 361–373. Reprinted in Saliba, G. 1994. *A History of Arabic Astronomy*, pp. 245–257. New York: New York University Press.

(1992). The Role of the Astrologer in Medieval Islamic Society. *Bulletin d'études orientales*. 44: 45–67.

(1994). *A History of Arabic Astronomy: Planetary Theories During the Golden Age of Islam*. New York: New York University Press.

Sankey, H. (1993). "Kuhn's Changing Concept of Incommensurability." *British Journal of the Philosophy of Science*. 44: 749–774.

(1998). "Taxonomic Incommensurability" *International Studies in the Philosophy of Science* 12, 1: 7–16.

(2008). *Scientific Realism and the Rationality of Science*. Aldershot: Ashgate.

(2012). "Methodological Incommensurability and Epistemic Relativism." *Topoi* 32: 33–41.

(forthcoming). "Kuhn, Coherentism and Perception." In P. Melogno, H. Miguel and L. Giri (eds.), *Perspectives On Kuhn*.

Sayili, A. (1960). *The Observatory in Islam and its Place in the General History of the Observatory*. Ankara: Türk Tarīh Kurumu Basimevi.

Scerri, E. R. (2007). *The Periodic Table: Its Story and Its Significance*. Oxford: Oxford University Press.

(2013). *A Tale of Seven Elements*, New York: Oxford University Press.

(2016). *A Tale of Seven Scientists and a New Philosophy of Science*, New York: Oxford University Press.

(2018). "Antonius van den Broek, Moseley and the Concept of Atomic Number." In R. Edgell, R. MacLeod, E. Bruton (eds.), *For Science, King and Country – Henry Moseley*, pp. 102–118. Uniform Press.

(2019). *The Periodic Table: A Very Short Introduction*, 2nd ed. Oxford: Oxford University Press.

(2020). *The Periodic Table, Its Story and Its Significance*, New York: Oxford University Press.

Scerri, E. R. and Ghibaudi, E. (eds.) (2020). *What is an Element*, New York: Oxford University Press.

Scheffler, I. (1967). *Science and Subjectivity*. Indianapolis: Hackett.

(1972). "Vision and Revolution: A Postscript on Kuhn." *Philosophy of Science* 39: 366–374.

Schilpp, P. (ed.) (1963). *The Philosophy of Rudolf Carnap*. La Salle: Open Court.

Schlick, M. (1931/1959) "The Turning Point in Philosophy." In A. Ayer (ed.), *Logical Positivism*, pp. 53–59. Chicago: Free Press.

Schofield, C. J. (1981). *Tychonic and Semi-Tychonic World Systems*. New York: Arno Press.

Schopenhauer, A. (1851/2014): "Sketch of a History of the Doctrine of the Ideal and the Real." In S. Roehr and C. Janaway (eds.). *Schopenhauer: Parerga and Paralipomena. Short Philosophical Essays*, vol. 1, pp. 7–30. Cambridge: Cambridge University Press.

Schrecker, E. (1986). *No Ivory Tower: McCarthyism and the Universities*. New York: Oxford University Press.

Seigel, K. (1951). "College Freedoms Being Stifled by Students' Fear of Red Label." *New York Times*, May 10.

Sellars, W. (1967). *Science and Metaphysics: Variations on Kantian Themes, The John Locke Lectures for 1965–66*. London: Routledge & Kegan Paul.

(2007). *In the Space of Reasons. Selected Essays of Wilfrid Sellars*, ed. K. Sharp and R. B. Brandom. Cambridge, MA: Harvard University Press.

Shan, Y. (2018). "Kuhn's 'Wrong Turning' and Legacy Today." *Synthese* online, 28 February.

Shapere, D. (1964). "The Structure of Scientific Revolutions." *Philosophical Review* 7: 383–394.

Sigurðsson, S. (1990). "The Nature of Scientific Knowledge." *Harvard Science Review*, Winter issue: 18–25.

(1990/2016). "The Nature of Scientific Knowledge: An Interview with Thomas S. Kuhn." In A. Blum, K. Gavroglu, C. Joas, and J. Renn (eds.), *Shifting Paradigms. Thomas S. Kuhn and the History of Science*, pp. 17–30. Proceedings 8, Max Planck Institute for the History of Science. Berlin: Edition Open Access.

Smith, A. D. (1999). "Primary-Secondary Distinction." In *Routledge Encyclopedia of Philosophy*, pp. 684–687 London: Routledge.

Stone, A. (2018). "Lewis and Cavell on Ordinary Language and Academic Philosophy." *Philosophical Inquiries* 6, 1: 75–96.

Strauss, L. (1954). "Remarks Prepared by Lewis L. Strauss, Chairman, United States Atomic Energy Commission, for Delivery at the Founder's Day Dinner, National Association of Science Writers," September 16, (available at www.nrc.gov).

Strawson, P. (1966/2018). *The Bounds of Sense*. London: Routledge.

Strong, R. (1973). *Splendor at Court: Renaissance Spectacle and the Theater of Power*. Boston: Houghton Mifflin.

Suppe, F. (ed.) (1974). *The Structure of Scientific Theories*. Chicago: The University of Illinois Press.

Swerdlow N. M. (1973). "The Derivation and First Draft of Copernicus's Planetary Theory." *Proceedings of the American Philosophical Society* 117: 423–512.

(2000). "Copernicus, Nicholas (1473–1543)." In W. Applebaum (ed.), *Encyclopedia of the Scientific Revolution from Copernicus to Newton*, pp. 162–168. New York: Garland.

Swerdlow, N. M. and Neugebauer, O. (1984). *Mathematical Astronomy in Copernicus's De Revolutionibus*. 2 vols. New York: Springer-Verlag.
THE. (2009). "Most Cited Authors of Books in the Humanities, 2007." *Times Higher Education*, 26 March 2009. www.timeshighereducation.co.uk/story.asp?sectioncode=26&storycode=405956, (accessed February 24, 2020).
Thorndike, L. (1949). *The Sphere of Sacrobosco and Its Commentators*. Chicago: University of Chicago Press.
Timmins, A. (2019). "Between History and Philosophy of Science." *HOPOS* 9, 2: 371–387.
Toomer, G. J. (1998). *Ptolemy's Almagest*. Princeton: Princeton University Press.
Toulmin, S. (1970). "Does the Distinction Between Normal and Revolutionary Science Hold Water?" In I. Lakatos and A. Musgrave (eds.), *Criticism and the Growth of Knowledge: Proceedings of the International Colloquium in the Philosophy of Science 1965, Volume 4*, pp. 39–47. Cambridge: Cambridge University Press.
  (1972). *Human Understanding: The Collective Use and Evolution of Concepts*, Princeton, NJ: Princeton University Press.
Tredwell, K. A. and Barker, P. (2004). "Copernicus' First Friends: Physical Copernicanism from 1543 to 1610." *Acta Philosophica/Filozofski vestnik* 25, 2: 143–166.
Tremaine, S. (2011). "John Norris Bahcall: 1934–2005." *Biographical Memoirs of the National Academy of Sciences*. National Academy of Science, Washington, DC.
Uebel, T. (2012). "De-synthesizing the Relativized A Priori." *Studies in the History and Philosophy of Science* 43: 7–17.
Ursus, N. R. (1588). *Nicolai Raymari Ursi dithmarsi Fundamentum astronomicum*. Strasbourg: Bernhardus Iobin.
  (2012). *Astronomischer Grund: Fundamentum Astronomicum 1588 des Nicolaus Reimers Ursus*, ed. and tr. C. Thierfelder and D. Launert. Frankfurt: Deutsch.
Van Dalen, B. (2007). "Ulugh Beg: Muḥammad Ṭaraghāy ibn Shāhrukh ibn Tīmūr." In T. Hockey et al. (eds.), *The Biographical Encyclopedia of Astronomers*, pp. 1157–1159. Springer Reference. New York: Springer, 2007.
Vesel, M. (2014). *Copernicus: Platonist Astronomer-Philosopher: Cosmic Order, the Movement of the Earth, and the Scientific Revolution*. Frankfurt am Main: Peter Lang.
Vesey, G. N. A. (1956). "Seeing and Seeing As." *Proceedings of the Aristotelian Society*, New Series, 56: 109–124.
Vision, G. (1982). "Primary and Secondary Qualities: An Essay in Epistemology." *Erkenntnis* 17: 135–169.
Vogt, T. (2017). "Book Review of Eric Scerri: A Tale of Seven Scientists and a New Philosophy of Science." *Hyle, International Review for the Philosophy of Chemistry* 23: 107–109.
Von Hoffman, N. (1988). *Citizen Cohn*. New York: Doubleday.
Voss, D. L. (1985). *Ibn al-Haytham's Doubts Concerning Ptolemy: A Translation and Commentary*. Ph.D. dissertation. University of Chicago.

Westman, R. S. (1975). "The Melanchthon Circle, Rheticus, and the Wittenberg Interpretation of the Copernican Theory." *Isis* 66, 2: 164–193.
　(1980). "The Astronomer's Role in the Sixteenth Century: A Preliminary Study." *History of Science* 18, 2: 105–147.
　(2011). *The Copernican Question: Prognostication, Skepticism and Celestial Order.* Berkeley: University of California Press.
　(2017). *Copernicus and the Astrologers: Dibner Library Lecture, December 12, 2013.* Washington, DC: Smithsonian Libraries.
Wilson, M. (2006). *Wandering Significance: An Essay on Conceptual Behavior.* Oxford: Oxford University Press.
Wilson, T. (2004). *Strangers to Ourselves: Discovering the Adaptive Unconscious.* Cambridge, MA: Harvard University Press.
Wimsatt, W. C. (2007). *Re-Engineering Philosophy for Limited Beings.* Chicago: University of Chicago Press.
Winsberg, E. (2010). *Science in the Age of Computer Simulation.* Chicago: University of Chicago Press.
Wittgenstein L. (1953/2009). *Philosophical Investigations*, tr. G. E. M. Anscombe, P. M. S. Hacker and J. Schulte, 4th ed. Blackwell, Oxford. Abbreviated as PI paragraph for what used to be Part I in previous editions and PPF (Philosophy of Psychology – A Fragment) paragraph for what used to be Part II.
　(1958). *The Blue and Brown Books.* New York: Harper Torchbooks.
　(1967/1970). *Zettel.* Berkeley, CA: The University of California Press.
　(1988). *Wittgenstein's Lectures on Philosophical Psychology 1946–1947*, ed. P. T. Geach. New York: Harvester Wheatsheaf.
　(1982/1990). *Last Writings on the Philosophy of Psychology*, vol, I, eds. G. E. M. Anscombe, G. H. von Wright, H. Nyman. Oxford: Basil Blackwell.
　(2005). *The Big Typescript.* Oxford: Blackwell Publishing.
WorldCat. (2020). "Du Chevreul, Jacques." http://worldcat.org/identities/lccn-no2017098045/
Wray. K. B. (2011). *Kuhn's Evolutionary Social Epistemology.* Cambridge: Cambridge University Press.
　(2015). "Kuhn's Social Epistemology and the Sociology of Science." In W. J. Devlin and A. Bokulich (eds.), *Kuhn's Structure of Scientific Revolutions – 50 Years On*, pp. 167–183. Dordrecht: Springer.
　(2016). "The Influence of James B. Conant on Kuhn's Structure of Scientific Revolutions." HOPOS 6, 1: 1–23.
　(2018a). "The Atomic Number Revolution in Chemistry: A Kuhnian Analysis, *Foundations of Chemistry* 20: 209–217.
　(2018b). "Thomas Kuhn and the T. S. Kuhn Archives at MIT." *OUPblog* (May 27, 2018). https://blog.oup.com/2018/05/thomas-kuhn-archives-mit/, (accessed March 4, 2020).
　(2019). "Kuhn, the History of Chemistry, and the Philosophy of Science." *HOPOS* 9, 1: 75–92.

Wray, K. B. and Andersen, L. E. (2019). "Reporting the Discovery of New Chemical Elements: Working in Different Worlds, Only 25 Years Apart." *Foundations of Chemistry*. (forthcoming).

Wray, K. B. and Bornmann, L. (2015). "Philosophy of Science Viewed through the Lense of 'Referenced Publication Years Spectroscopy' (RPYS)." *Scientometrics* 102: 1987–1996.

Yates, F. C. (1947). *The French Academies of the Sixteenth Century*. Warburg Institute Studies, v. 15. London: Warburg Institute, University of London.

Zhang, Z. X. (2008). "Research on the Origin and Spread of the Star Catalogue *Xiyang Xinfa Lishu*." *Journal of Dialectics of Nature* 30, 2: 81–86.

Zhang, F. Kearns, S. L., Orr, P. J., Benton, M. J., Zhou, Z., Johnson, D., Xu, X., and Wang, X. (2010). "Fossilized Melanosomes and the Colour of Cretaceous Dinosaurs and Birds." *Nature* 463: 1075.

Index

a priori, 19, 27, 32–42, 79, 198
algorithm, 107, 159, 186, 189, 196, 231
Almagest, 147, 156, 160, 164, 167
anomaly, 91, 94, 96, 99, 127–129, 133–134, 138, 148–149, 160–162, 175, 179, 183–184, 188, 210, 228, 232–234
apparent motion, 12–13, 16–18, 24–25, 156
Archimedean platform, 118–119, 122–123, 221
Aristotle/Aristotelian physics, 15, 31–32, 34, 77, 80, 107, 109, 118, 147, 152, 155–156, 159, 162–163, 166–168, 174, 181, 210, 216
art, 4, 65, 72–79, 80, 181, 203
aspect seeing, 169–170, 173, 175, 178, 180, 182–183
atomic number, 4, 126–140
atomic weight, 4, 93, 126–139
Averroes/Averroist, 156–159

Bernal, John Desmond, 55, 56
biological speciation, 116, 136, 198, 211, 214, 233
Bohr, Niels, 80, 126–127
Boyle, Robert, 15, 17, 32, 59
Brahe, Tycho/Tychonic astronomy, 157–158, 159, 163, 164, 166–168
Bruner, Jerome and Postman, Leo, 175, 228

Carnap, Rudolf, 28, 40–41, 66–72, 73, 79–81
categories, 23, 37, 40, 114–115, 118, 163, 176, 177, 179, 198, 228
Catholic, 158, 166
Cavell, Stanley, 36, 173, 179
Chang, Hasok, 29, 222–228, 232, 234, 236
chemical element. See element, chemical
Chemical Revolution, 89, 135, 163
chemistry, 4, 29, 47, 49, 58–59, 80, 125–127, 131–134, 135–136, 195, 208–209
classes, 195–196, 229, See also kinds (as related to lexicons)
classical mechanics, 32, 126, 128, 193–194, 195, 197–200, 208

coherence/incoherence, 11, 43, 49, 101, 124, 203–204, 208, 209–210, 211, 212, 213, 216–218, 220–221, 224
Conant, James B., 4, 28, 31, 32, 38, 47–54, 56–57, 58–63, 66, 78–80, 85, 173
concept-laden observation. See theory-laden observation
conceptual change, 30, 163, 180, 229
conceptual scheme/framework, 33, 51, 114, 199, 226
consensus, 3, 48, 61–62, 88–90, 92, 96–97, 100, 146, 184, 222–223, 226, 228, 230–232, 234, 236
continuity/discontinuity, 73, 79, 80, 87, 125–127, 131, 145, 151–153, 155, 158, 161, 165–166, 168, 176, 197, 199–200, 206, 208, 209, 211–212, 232, 235
convergent (on the truth), 205–206
convergent thinking, 92, 95, 183
conversion, 51, 64, 180, 181–182, 230
Copernicus, N./Copernican astronomy, 3, 5, 9–26, 115, 137, 138, 145–146, 148, 152–168, 220, 229, 232
correspondence theory of truth, 4, 105, 106, 116, 121, 124, 204, 206, 207, 212, 215, 217–218, 219, 221
creativity, 50, 63, 79, 86, 169, 180–184, 190, 192–193, 198, 231, 234, 236
crisis, 41, 54, 89, 91, 96, 107, 128–129, 130, 133, 134, 145, 160, 162, 168, 180, 187, 190–192, 227, 232, 234
cumulative/noncumulative progress, 38, 66–67, 71–74, 77–78, 81, 86, 94, 107, 168, 180, 185, 197–198, 226
cyclical model (Kuhn's), 4, 160, 166

Darwin, Charles/Darwinian theory, 103, 114, 118, 204–205, 212, 214
Descartes, Rene/Cartesian philosophy, 15, 19, 31, 35, 37, 168, 174, 177–178
Dewey, John, 29, 218
disciplinary matrix, 98–101, 104, 107, 183

264

Index

discovery, 3, 5, 15, 25, 50, 54, 72, 75, 126, 128–129, 130, 132, 133–135, 139–140, 167, 183, 185–190, 194–197, 199–201, 215
divergent thinking, 38, 95–96, 110, 183
duck/rabbit, 5, 169–172, 175, 178–180

Einstein, Albert, 22, 23, 73, 80, 82, 103, 108, 186, 192, 194, 199
element, chemical, 4, 93, 126–129, 130–140, 163, 229
evolution, biological, 204–205, 208, 214, 221, 233
evolutionary dimensions (in Kuhn's philosophy), 3, 5, 106, 202–212, 213–215, 217, 221
exemplar, 4, 50, 98–102, 104, 182, 187–192, 193, 194–196, 198–199, 205, 226, 228, 231, 234

Feyerabend, Paul, 106, 166, 224, 225
fragmentation (of science), 90, 202–203, 209–211, 219, 221
Frank, Philipp, 28, 31, 37–38, 61–62, 63
Friedman, Michael, 28, 41–42

Galilei, Galileo, 13, 15, 32, 50, 79, 146, 156–157, 159, 163, 168, 174, 196, 198, 232
general education, 32, 38, 52–53, 56, 59, 61, 62
geocentric (cosmology), 10–11, 13, 145, 164, 165, 167
gestalt figure/switch/psychology, 5, 29, 36, 108, 170, 172, 179–182, 188, 194, 200, 230
Gombrich, E., 75

Hanson, N. R., 46, 108, 170, 171–172, 179, 181, 195
heliocentric (cosmology), 13, 145, 157, 163–166
Hempel, C., 66, 175, 193
historical development (of science), 10, 13, 16, 19, 73, 78, 85–86, 208, 209–210, 215, 217
historical perspective, 207, 211, 215
historical school (of philosophy of science), 212, 223
historiography of science, 21, 44, 155, 212
history and philosophy of science, 2, 13, 18, 25, 41, 45, 54, 63, 85, 125, 186, 193, 202, 205
history of art, 72, 75, 80
history of philosophy, 19, 43, 70, 71, 79–80, 106, 188
history of science, 26, 31–32, 34, 41–44, 45, 47, 49, 52, 54, 58–60, 61, 63–64, 68, 70–71, 76, 78, 85, 87, 88–89, 96, 100, 103, 119, 131, 133–134, 145–146, 152, 158, 160, 176, 180, 182, 186, 188, 196–197, 200, 203, 206, 208, 214, 221, 223, 226, 233
Hoyningen-Huene, Paul, 2, 3, 11, 22, 24, 32, 33, 35–36, 80, 109, 116, 160, 176–177, 202, 213, 214, 224
Hume, David, 19–20, 31, 35, 37

idealism, 9, 11, 30, 40, 79, 80, 118, 177
incommensurability, 3–4, 40, 72, 74–76, 80–81, 89, 105–124, 129, 131, 137, 145, 166, 168, 197, 198–199, 200, 209, 210–211, 217, 230
indeterminacy of translation, 102, 106, 110–111
innovation, 15, 21, 61, 91, 98, 146, 155, 180, 183–184, 185–186, 187, 191, 195, 198, 199, 200
interpretation, 98, 100, 109, 112–114, 119, 131, 136, 170, 172–175, 178, 185, 191–192, 199, 231–232, 234
isotopes, 129, 130, 134, 136, 138–139

Jesuit, 159, 166, 168
Jupiter, 138, 147, 159, 167

Kant, Immanuel, 3–4, 10, 19–21, 23–24, 27–44, 77, 79–80, 114, 118, 177, 198, 213, 216
Kepler, Johannes, 145–146, 156, 158–159, 163, 165–168, 232
kinds (as related to lexicons), 113, 115, 138, 229, 233, See also classes
Koyré, Alexandre, 29, 32, 34, 81
Kuhn Archives, Thomas S. See TSK Archives

Langer, Susanne, 28–31, 35–36
Laudan, Larry, 225, 227, 234
Lavoisier, A., 135, 163
lexical change, 4, 118, 133, 137, 209, 211, 229
lexicon, 26, 105, 112, 113–124, 129–130, 136–140, 216, 217, 219, 221, 226, 228–230, 233–234, 236
lexicon-laden observation. See theory-laden observation
local incommensurability, 105–106, 111–113, 115–123
Locke, John, 15, 58, 79
logic, 5, 20, 32–34, 36, 49, 61, 70, 98, 101–104, 123, 152, 162, 169, 172, 181, 182–183, 185–187, 189, 194–195, 198–199, 201, 216, 218, 220
logical empiricism/logical positivism, 27, 37–38, 41, 43, 65–69, 70, 71–72, 76–77, 78–82, 132, 186
Lowell Lectures, The Quest for Physical Theory, 32, 54, 56, 58, 62, 82
Lutheran, 157, 166

Marcum, James, 36, 74, 106, 124, 202, 208, 214
Mars, 138, 147, 157, 167
Masterman, Margaret, 32, 33, 97–98, 104, 226
McCarthy, Senator Joseph/McCarthyism, 51–55, 57, 62, 64
meaning change, 71, 106, 108, 109, 110–113, 115, 121, 130, 132, 135, 137, 173, 182, 191, 194, 198–199
Mendeleev, D. I., 128, 133, 135–136, 137
Mercury, 128, 138, 147, 148, 151–152, 153–155, 159, 160–161, 163, 167

metaphysics, 3, 9–11, 15–17, 19, 22–25, 42, 68, 78, 81, 92, 99, 104, 109, 116, 123, 152, 191, 218
methodological incommensurability, 105–107, 117, 209
Mill, John Stuart, 58, 225
mind-independent world, 11, 105, 109, 116–119, 122–124, 178, 204, 206, 207, 213, 215
model, 19, 33, 42, 57, 61, 65, 67, 71, 75, 79, 81, 97–104, 131, 133, 148–156, 158, 160–162, 165, 166, 181, 182, 190, 192–193, 195–197, 198–199, 201, 203, 205, 207, 209, 217, 221, 227
mop-up work, 4, 87, 91, 181

Nagel, E., 66–67, 193
National Science Foundation (NSF), 56–57
Neo-Kantian(ism), 21, 27–31, 33, 35, 41, 199
Neurath, Otto, 55–56, 68–69, 72, 80
Newton, Isaac/Newtonian physics, 13, 15, 22, 32, 37, 50, 59, 73, 79, 82, 93, 103, 107, 118, 128, 145, 157, 158–159, 166, 168, 191, 193–194
niche, 118, 177–178, 203–204, 211, 213–216, 217
no-overlap principle, 113, 115, 137–138, 140, 211, 229
normal science/normal research, 3–5, 35, 39, 41, 61, 62, 78, 85–88, 90–104, 125, 133, 146–149, 152–153, 160–162, 168, 175, 180, 187–192, 194–195, 197, 199, 222, 226–227, 230–231, 236
novelty, 9, 85, 90–92, 94–96, 100–103, 147, 161, 163, 166, 168, 169, 180, 183, 232

object-sided, 11–15, 16–24
oxygen, 32, 50, 134, 135, 163, 174, 185, 186

paradigm, 4–5, 9, 25, 28–30, 32–34, 36, 38–39, 40–42, 62, 64, 75–76, 80, 90, 96–99, 100, 102, 105, 107–109, 115, 117–118, 119, 121–123, 129–130, 132–133, 136–137, 146–147, 160–162, 168, 170, 172, 174–177, 178–180, 182–184, 187–193, 199, 216, 226, 228, 231–233, 236
pedagogy, 30, 49, 63, 88, 100
periodic table of elements, 127–129, 131, 133, 134–136, 235
phlogiston/phlogiston theory, 32, 163, 223–224
Piaget, Jean, 29, 31, 33–34, 36, 40, 42
Plato/Platonist, 11, 13, 15, 164–165
pluralism, 5, 48, 217, 223–228, 231–232, 233–236
Plurality of Worlds, xiii, 1, 116, 117, 124, 202, 203, 212, 214, 215, 218, 219
Popper, Karl R., 70, 73, 75, 78, 81, 86, 88, 90, 100, 186, 227, 231, 234
positivism. See logical empiricism/logical positivism
printing, 159, 164
progress (scientific), 18, 33, 51, 59, 63, 66–68, 71–73, 76, 77–78, 87–90, 92, 95, 97, 98, 103, 148, 179, 181–183, 185, 188, 197, 203, 204, 206, 208, 214, 221, 223, 229
Ptolemy/Ptolemaic astronomy, 115, 137, 138, 139, 145–152, 156–163, 165–167, 229
puzzle-solving tradition, 36, 43, 50, 88–90, 91–102, 149, 187, 189–192, 195–196, 231

quantum mechanics, 10, 22, 23, 32, 41, 49, 89, 126–127, 131, 159, 186, 191–194, 196–200, 203, 230

Regiomontanus, 156, 164–165
Reichenbach, H., 30, 38, 41, 68, 70–72, 79–80
relativity theory, 10, 22, 32, 41, 159, 193–194, 195, 199
revolution/revolutionary change, 3–5, 10, 15, 17–19, 23, 28, 61, 66, 75–76, 78, 80, 89, 107–108, 109, 111, 117, 125–141, 151, 158, 162–163, 168, 176, 178–183, 187–188, 190–192, 193–200, 202, 204, 208–211, 229–230, 232
Richardson, Alan, 32, 35, 41, 43–44, 66, 67
Ruphy, Stephanie, 224, 226

Sarton, G., 66, 73, 78, 80
Saturn, 138, 147, 167
Schlick, M., 69, 72, 80
scientific community/community (of scientists), 4, 32, 37, 39, 43–44, 50, 60–62, 63, 76, 82, 87–90, 95–102, 104, 107, 109, 114–115, 118, 120–121, 127–128, 140, 176, 180, 182, 183, 185, 211, 215, 218, 222, 225, 228, 232–234, 236
scientific method, 49, 58, 60, 186, 222
scientific philosophy, 40–41, 66, 67–72, 77, 79–82
scientific revolution, 3–5, 26, 28, 45, 64, 75, 107, 111, 125–132, 136–138, 140–141, 151, 160, 163, 169–170, 181, 188, 197, 198, 202, 209, 229
Scientific Revolution, the, 17–18
scientific specialty, 68, 116, 118, 136–137, 198, 208–211, 213, 217, 221, 222, 225, 228–231, 233–236
'seeing' and 'seeing as', 169–173, 179
semantic incommensurability, 105–108, 110–112, 117, 119–123, 209
similarity relation, 36, 114, 187, 196, 199
social structure of science, 82, 104
sociology of science, 61–62, 63, 188, 189
specialization, 68, 116, 198, 208, 209–211, 235–236, See also scientific specialty
speciation, 137, 206, 208–211, See also biological speciation
Strong Programme, the, 231
subject-sided, 3, 11–15, 16, 18–24, 26
Sun, the, 10–14, 17–19, 94, 115, 137–138, 139, 145, 147–148, 157–159, 162–163, 165–167, 229

symbolic generalization(s), 98–100, 102
systems of practice, 224, 227, 232, 234–236

taxonomy/taxonomic categories, 113–115, 120, 137, 138, 198, 209–211, 228
Thalheimer Lectures, 29, 82
theoretical change, 66, 126–127, 128, 131, See also revolution/revolutionary change
theoretical monism, 3, 5, 222–223, 225–228, 232, 234–235, 236
theoretical pluralism, 224–225, 227
theoretical terms/notions, 36, 131, 229–230
theory of truth, correspondence. See correspondence theory of truth
theory-dependence (of observation). See theory-laden observation
theory-laden observation, 3, 5, 108–109, 114, 117, 121–123, 169, 171, 179
thought experiment, 50, 80, 81–82
true motion, 12–13, 15, 16, 18

truth, 5, 13–15, 21, 38, 41, 50–51, 70–71, 116–124, 184, 185, 188, 202–208, 211, 212–221, 225, See also correspondence theory of truth
TSK Archives, xiii–xiv, 1, 36, 45, 46, 53, 54, 58, 62, 63, 73–76, 82

unified world view/unified science/unity of knowledge, 68, 204, 206, 209–210, 216

Venus, 138, 147, 159, 163, 167

Whig/anti-whiggish history, 17, 186, 188, 193
Wittgenstein, Ludwig, 28, 30–31, 33, 35–36, 39, 169–175, 179, 189
world change, 10, 25–26, 169, 175, 176, 230
'world-in-itself' vs. phenomenal world, 176–178, 213
Wray, K. Brad, 2, 4, 5, 10, 28–29, 32, 61, 82, 125–141, 198, 202, 207, 209, 211, 214, 216, 226, 228, 229, 231, 233

X-ray, 32, 50, 129, 134, 171–172

www.ingramcontent.com/pod-product-compliance
Ingram Content Group UK Ltd.
Pitfield, Milton Keynes, MK11 3LW, UK
UKHW022001190125
453752UK00007B/50